Conceptual Agricultural Microbiology

Conceptual Agricultural Microbiology

Yogesh Kumar
R. K. Naresh
P.K. Singh
Mukesh Kumar
S. P. Singh
Deepak Chaudhari
Himanshu Tiwari

CRC Press
Taylor & Francis Group
Boca Raton London New York

CRC Press is an imprint of the
Taylor & Francis Group, an **informa** business

-EPH-

First published 2025
by CRC Press
4 Park Square, Milton Park, Abingdon, Oxon, OX14 4RN

and by CRC Press
2385 NW Executive Center Drive, Suite 320, Boca Raton FL 33431

CRC Press is an imprint of Informa UK Limited

British Library Cataloguing-in-Publication Data
A catalogue record for this book is available from the British Library

Print edition not for sale in South Asia (India, Sri Lanka, Nepal, Bangladesh, Pakistan or Bhutan).

ISBN13: 9781032788111 (hbk)
ISBN13: 9781032788418 (pbk)
ISBN13: 9781003489474 (ebk)

DOI: 10.1201/9781003489474

Typeset in Adobe Caslon Pro
by Elite Publishing House, Delhi

-EPH-

Contents

Practical Part-II

Preface

Microbiology is the science of small living organisms and deals with the wide occurrence of such organisms in soil, water and air. Being too small to be seen with unaided eyes, these organisms are termed as microorganisms. Microorganisms refer to express viruses, bacteria, fungi, protozoa and some algae. Microorganisms are recognized as integral and functionally important components of diverse habitats, ranging from soil collectives to the human surroundings. Virtually microorganisms are omnipresent and are imposed on all aspects of life; therefore the science of microbiology has assumed a central position of great significance in biological science as whole. Conventionally, in elementary cycle of the nature, microorganisms used to be described only as decomposer whereas green plants as producers and animals as consumers. However, with the advances in our biological knowledge base, it has been realized that plants, animals and microorganisms are interdependent in many sense.

Among the biological sciences, microbiology is well established and related with various fields in the present era. Agricultural microbiology is a blooming research field emerging from the intersection of general microbiology and microbial ecology to the agricultural biotechnologies. The ultimate aim of agricultural microbiology is a wide-ranging study of beneficial bacteria and fungi interacting with agriculturally important plants, thereby meeting the global demand of food in an eco-friendly manner. Our aim is to torch the current status and application of microbiology in modern agriculture. Modern soil microbiology represents an integration of microbiology with the concepts of soil science, chemistry, and ecology to understand the functions of microorganisms in the soil environment. The surface layers of soil contain the highest numbers and variety of microorganisms, because these layers receive the largest amounts of potential food sources from plants and animals. The soil biota form a belowground system based on the energy and nutrients that they receive from the decomposition of plant and animal tissues. The primary decomposers are the bacteria and fungi. Thus in a broad sense, the book will be useful to the students of Microbiology. Botany, Biochemistry and Agricultural Microbiology, both at the under-graduate and post-graduate level.

Microorganisms of various types exist in all three domains of life and they are by far the most abundant life forms on Earth. Microscopic biological agents include bacteria, archaea, protists, fungi, parasitic worms and viruses. Although a small percentage of microorganisms are harmful to certain plants and animals and may cause serious disease in humans, the vast majority of microorganisms provide beneficial services, such as assisting in water purification and the production of certain foods, and many are essential for the proper functioning of Earth's ecosystems.

Reasonable efforts have been made to publish reliable data and information but the author/s, editor/s and publisher cannot assume responsibility for the validity of all materials or the consequences of their use. The author/s, editor/s and publisher have attempted to trace and acknowledge the copyright holders of all material reproduced in this publication and apologize to copyright holders if permission and acknowledgements to publish in this form have not been taken. If any copyright material has not been acknowledged please write and let us know so we may rectify it, in subsequent reprints. We are sure this book is ideally designed for farmers, agriculturalists, product managers, farm holders, manufactures, governmental professionals, researchers, academicians, and students with zeal and enthusiasm.

Authors

About the Authors

Dr. Yogesh Kumar is working as an Associate professor in the Department of Soil Science and Agricultural Chemistry, Sardar Vallabh bhai Patel University of Agriculture & Technology, Meerut , (U.P.) – India. Dr. Kumar has more than 19 years experience of teaching, Research and Extension with specialization in Soil Chemistry and Soil biology. He has guided 15 M.Sc Ag., students. Dr. Kumar has published 04 Books. He is a prolific writer and has published more than 45 research papers referred in national and international journals and 50 popular articles in different rural development magazines for the welfare of farmers. Dr. Kumar is the recipient of four awards from different national organizations and societies in the national and international events. He is life member of various professional societies. He is the reviewer of various national and International Journals. Dr. Kumar handles two research project funded by various prestigious national bodies of India as Principal Investigator and Co- principal Investigator. Currently He is working as PI "Forecasting Agriculture output using space, Agro meteorology and Land based observation (FASAL) which is funded by Indian Meteorology Department, Ministry of Earth Science, Govt., of India.

Dr. R. K. Naresh is a well known professor of Agronomy and he has professional experience of more 33 years of Teaching, Research and Extension with specialization of Conservation Agriculture, Crop Management and Water Management at Sardar Vallabhbhai Patel University of Agriculture &Technology, Meerut-250110 (U.P.) - India. He has guided more than eleven Ph. D and fifteen M.Sc. students in their research programme. He is a prolific writer and has published several manuals for teaching different courses, a lot of chapters in scientific books, more than 300 research papers published in referred national and international Journals and many popular articles in different rural development magazines for the welfare of farmers.

He has also authored twenty six books (Irrigation of Crops, Theory, Principles and Practices; Water Footprint on Agriculture; Agriculture and Climate Change; Research Trends in Agriculture Sciences; Advances in Agriculture Sciences, Crop Production and Management; Agroforestry and its Applications; Precision Farming Geoinformatics and Nanotechnology; Techniques for Sustainable Crop Production; Rainfed Agriculture and Watershed Management; Introductory Agrometeorology and Climate Change; Cropping System and Sustainable Agriculture; Text Book on Crop Ecology; Principles and Practices of Weed Management; Principles of Organic Farming; Principles of Agronomy; Artificial Intelligence in Indian Agriculture; Soil Conservation and Watershed Management; Tropical and Sub- Tropical Fruits; Text Book On Orchard and Estate Management; Irrigation Management Theory, Principles and Practices; Soil Biology and Soil Health; Growth and Development of Horticultural Crops; Textbook On Agribusiness and Rural Marketing, Artificial Intelligence in Agriculture Opportunity, Risk and Future Perspectives). He has chaired a number of scientific sessions in different symposium, conferences and workshops and made recommendations to government for adoption and implementation.

He also worked as sub-Coordinator of the prestigious World Bank funded NATP Project and also led the production system research program a of Rice-Wheat Consortium (RWC) convened by CIMMYT,

Mexico and led by the 4 regional NARS (India, Bangladesh, Nepal and Pakistan). RWC programs on Farmer Participatory Adaptive Research conducted on small-farm holders field and won very prestigious

CG Awards, namely the King Baudouin Award 2004 and the CGIAR Excellence in Scientific Partnerships Award 2002 and the Manthan Award of the Government of India for developing OSCAR, an online App for weed identification and their management. Together with the Central Soil Salinity Research Institute, Dr. Naresh's researches led to (i) development of new irrigation water quality criteria for monsoonal climates of South Asia, (ii) Introduction of laser assisted precision land leveling system for efficient irrigation water management, and (iii) Initiate a human resource development program in the newly established SVPUA&T Meerut on Farmer Participatory Adaptive Research on Resource Conserving Technology (now better known as Conservation agriculture). With NATP and CG funds, Dr Naresh was able to promote the adoption of CA practices by farmers across several districts of the Western Uttar Pradesh. Benefits of the resource conserving technologies (Zero-tillage, reduced tillage and raised bed planting, surface seeding and DSR technologies are being realized by several hundred thousand farmers cultivating more than a million hectares which save on water, labor and fuel). Because of our concerted team efforts, the western UP has more than 300 Laser land leveling system service providers. He has published over 300 research articles in national and International journals and has carried out pioneering work for adapting conservation agriculture in wheat based sugar cane and RW-systems.

Dr. P.K. Singh is a well known Agronomist and he has professional experience of more 27 years of Extension with specialization of Sustainable Agriculture, Crop Management and Nutrient Cycling at Sardar Vallabhbhai Patel University of Agriculture & Technology, Meerut (U.P.), India. He is a prolific writer and has published several manulas / bulletins for vocational training and a lot of chapters in scientific books, research papers published in referred national and international Journals and many popular articles in different rural development magazines for the welfare of farmers.

He honoured with 12 awards at state and national level, and reviewed SREP of state Rajasthan, Tamil Nadu, M.P, Chhattisgarh, Bihar and U P. Dr Singh also handled Development of Climate Resilient Sustainable Integrated Farming System Model for Optimizing Farm Productivity for western U.P. farmers. He also holding the offices of 6 professional national level societies and also acting as editorial board members of three journals. He authored more than 30 scientific papers, 10 bulletins, 02 books, 15 book chapters and 150 extension articles for the benefit of the scientific and farming community. He has chaired a number of scientific sessions in different symposium, conferences and workshops and made recommendations to government for adoption and implementation. He is regularly called as resource person to deliver invited talks in seminars, symposiums and conferences. He acted as member of organizing committee of many seminars and symposiums.

Dr. Mukesh Kumar is working as an Associate Professor in the Department of Biotechnology, S V P University of Agri. & Tech., Meerut (UP) India. His Specialization is in Genetic Engineering for stress management and Plant Tissue Culture. He has more than nineteen years experience for teaching and Research of under Graduate and Post Graduate class. Author published more than fifty research papers in repute national and international journals. He has presented papers in India and Abroad and also chair the various sessions. Author received many awards and recognition in India and Abroad. Dr. Kumar organized various events in India and abroad. Author guided more than fifty students for master and doctoral degree and also submitted more than thirty gene sequences in NCBI, Gen Bank from various crops. Author holds various administrative and non-administrative posts. Dr Kumar received many awards from various organizations in national and international conferences. He is Life Member of various Society and organizations. Author handled more than five Projects as Principal Investigator and Co-Principal Investigator funded by various funding agencies of government.

Dr. S.P.Singh is a well known Associate professor of Soil science and Agricultural Chemistry and he has professional experience of more 18 years of Teaching, Research and Extension with specialization of Soil Fertility and Nutrient Management at

Sardar Vallabhbhai Patel University of Agriculture &Technology, Meerut-250110 (U.P.) - India. He has guided one Ph. D and nine M.Sc. students in their research programme. He is a prolific writer and has published more than 85 research papers published in referred national and international Journals and many popular articles in different rural development magazines for the welfare of farmers.

He has chaired a number of scientific sessions in different symposium, conferences and workshops and made recommendations to government for adoption and implementation. He is regularly called as resource person to deliver invited talks in seminars, symposiums and conferences. He acted as member of organizing committee of many seminars and symposiums. He is life member of three professional societies and annual member of two professional societies.

Deepak Chaudhari obtained his B.Sc. Agriculture (Hon.) Degree from Chattrapati Sahu Ji Mahraj University Kanpur and M.Sc. Agriculture (Soil Science and Agricultural Chemistry) from Sardar Vallabhbhai Patel University of Agriculture And Technology, Meerut, Uttar Pradesh, India. He has been awarded in many extra curriculum activities in SVPUAT, Meerut. He has attended many a seminars, webinars and workshops related to Agriculture. Presently he is working as Soil Testing Analyst in Uttam Sugar Mills Limited, Barkatpur –Bijnor (Uttar Pradesh).

Himanshu Tiwari obtained his graduated in Agriculture in 2018 from Acharya Narendra Deva University of Agriculture & Technology, Ayodhya (U.P.).He had obtained his Master's in Agronomy from Acharya Narendra Deva University of Agriculture & Technology, Ayodhya (U.P.) in 2020 and currently enrolled as Ph.D. Scholar (Department of Agronomy) in Sardar Vallabhbhai Patel University of Agriculture & Technology, Meerut (U.P.).He is a recipient of U. P. Mandi Parishad Scholarship. He has published more than 35 popular scientific articles, 25 research/review paper reputed journals. He has published 1 book and contributed to the books with 6 chapters. He has also published 1 patent on title "A liquid biodegradable fertilizer material made of biological components. He has received Best Master's Thesis Awards jointly organized by ICAR-IGFRI, ICAR-NAHEP, Research Station, Srinagar, Birsa Agricultural University and National Agriculture Development Cooperative Ltd. Baramulla, J & K. He has attended several national and international conferences and training. He is enrolled as a life member of Association of Agrometeorologists.

Chapter - 1

Introduction

Microbiology

The study of microscopic organisms that exist as single cells or clusters is called microbiology. The microorganisms include bacteria, viruses, archaea, fungi, and protozoa. In the microbial world, microorganisms are everywhere. Almost every natural surface is colonised by microbes. Some microorganisms can live quite happily in boiling hot springs, whereas others form complex microbial communities in frozen sea ice.

Most microorganisms are harmless to humans. We swallow millions of microbes every day with no ill effects. In fact, we are dependent on microbes to help us digest our food and to protect our bodies from pathogens. Microbes also keep the biosphere running by carrying out essential functions such as the decomposition of dead animals and plants bodies.

Microbes are the dominant form of life on planet Earth. More than half the biomass on Earth consists of microorganisms, whereas animals constitute only 15% of the mass of living organisms on Earth.

The Microbiology deals with

» How and where they live
» Their structure
» How they derive food
» Reproduction system
» Relation with plant and animals

The microorganisms can be divided into two distinct groups based on th
nucleus structure as-

Prokaryotes: The organism lacking true nucleus (membrane enclosed chromosom and nucleolus) and other organelles like mitochondria, Golgi body, endoplasmi reticulum etc. are referred as Prokaryotes (Exp. Bacteria and Archaea).

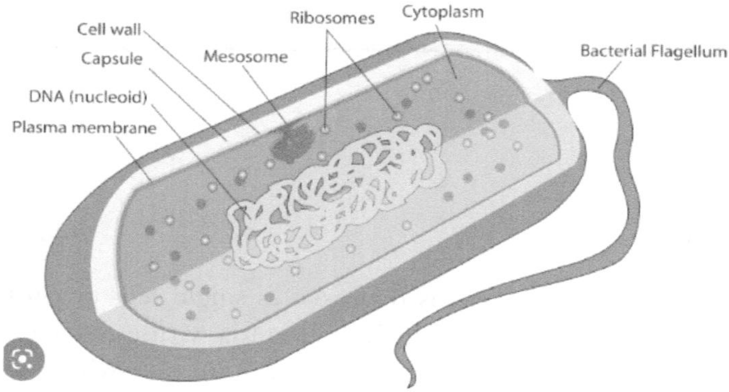

Fig. 1.1 Prokaryotes

Eukaryotes: The organism possessing membrance enclosed mucleus and other cel organelles are referred as Eukaryotes (Exp. Algae, fungi and protozoa).

Fig. 1.2: Algae

Fig. 1.3 :Algae

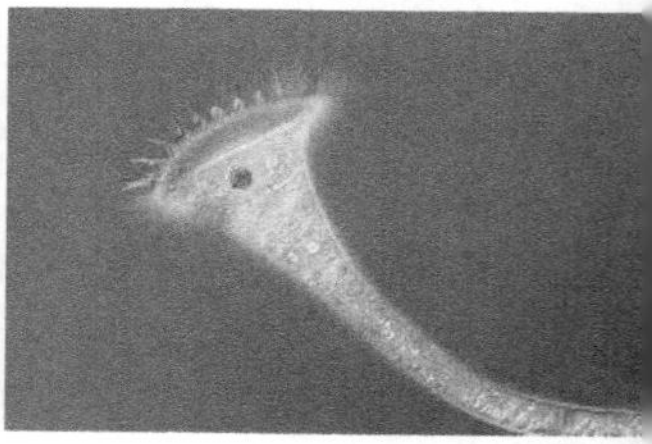

Fig. 1.4 :Algae

The microorganisms were divided into six distinct groups based on the phylogenetic, morphological, and physiological characteristics.

The microorganisms groups' are-

- Bacteria
- Archaea
- Fungi
- Algae

  - Protozoa
  - Viruses
- **Bacteria** are phylogenetically related group of unicellar prokaryotic organisms distinct from archeae.
- **Archaea** are phylogenetically related group of prokaryotes which are primitive and distinct from bacteria.
- **Fungi** are group of eukaryotic organisms lack of chlorophyll. They range in size and shape from single celled yeast to multicellular mushrooms.
- **An alga refers** the group of eukaryotic organisms with chlorophyll. They range in size and shape from single celled algae (Ex: Chlorella) to complex cellular structured plant like algae (Ex. Kelp).
- **Protozoa** are group of eukaryotic organisms lack of cell wall. The morphology, nutrition and physiology are different from other groups.
- **Viruses** are group of non-cellular organisms, parasite or pathogen to plant, animals and other micro-organisms. They are too small and can be visualized only under electron microscopes.

Application for Microbiology

- The microorganisms are a research tool to understand the chemical and physical basis of life.
- They are tools to analyse the biochemical and genetic background of living things.
- They are an excellent model for understanding the cell functions.
- They play an important role in the fields of medicine, agriculture, and industry.

Branches of Microbiology

Medical Microbiology: It deals with the microorganisms associated with the human body, diseases, diagnostic procedures, and preventive measures.

Aquatic microbiology: It deals with water purification, microbiological examination, biological degradation of wastes, and ecology.

Aero microbiology: It deals with the microbes in the air, contaminations, spoilage, and dissemination of diseases.

Food microbiology: It deals with the microorganisms associated with foods, foodborne pathogens, food spoilage, food preservation, and food safety.

Agricultural Microbiology: It deals with plant-associated microbes, soil fertility, plant and animal diseases, microbial degradation of organic matter, and soil nutrient transformation.

Industrial microbiology: It deals with the production of antibiotics and vaccines using microorganisms; fermentation and fermented products such as beverages, bread, and milk products; and the production of protein and hormones by genetically modified organisms.

Exomicrobiology: It exploitation for life in outer space

Geochemical microbiology: It deals with coal, gas, and mineral formation; mineral recovery through microbes from ore

Chapter - 2

Story of Microbiology

Obviously humans have had to deal with microbes even before recorded history. The first record of humans using microorganisms comes from ancient tablets from the Middle East.

Babylonians were using yeast to make beer over 8000 years ago and acetic acid bacteria to make vinegar over 6000 years ago. About 5000 years ago, the Persia (now Iran) region recorded the winemaking.

The Romans had gods for specific microorganisms. The Roman gods of mould and mildew were "Robigus" and "Robigo, which mean crop rust. Rust is one of the plant diseases caused by fungus. God Robigus was very much feared because of crop loss. About 2000 years ago, Romans proposed that diseases were caused by tiny animals. But fundamentalist religions had a strong hold over the progress. The real microbiology history starts from the 1600s, when people began to make crude lenses and microscopes.

The following are the scientists and their role in the development of microbiology:

Discovery of Microscope

Robert Hooke (1635-1700): One maker of microscope (30 x magnification) discovered the cell (Latin: cellulae meaning small room) from cross sections of cark. He noticed some microscopic fungi too.

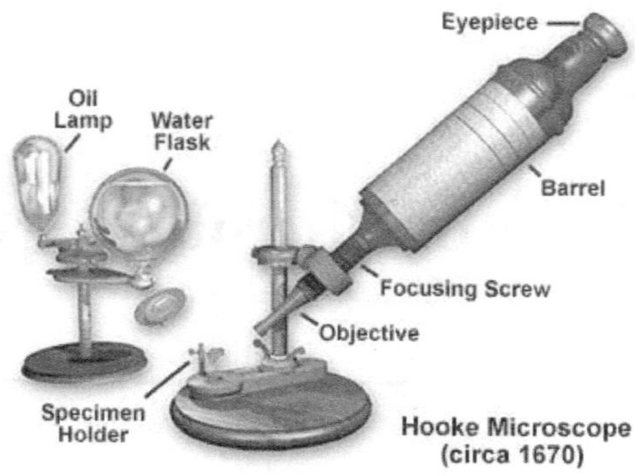

Fig. 2.1: Robert Hooke's Microscope

Antony Van Leeuwenhock (1632 – 1723): Lived during the same age as Robert Hooke, he was an amateur lens and microscope maker. His microscope can magnify 300x times. He was the first to describe the protozoa and bacteria. He observed some bacteria from plagues of his own teeth. He named them as animalcules. He was way ahead of hs time, and many scientists too did not believe him. But now, he is recognised as the Father of Microbiology.

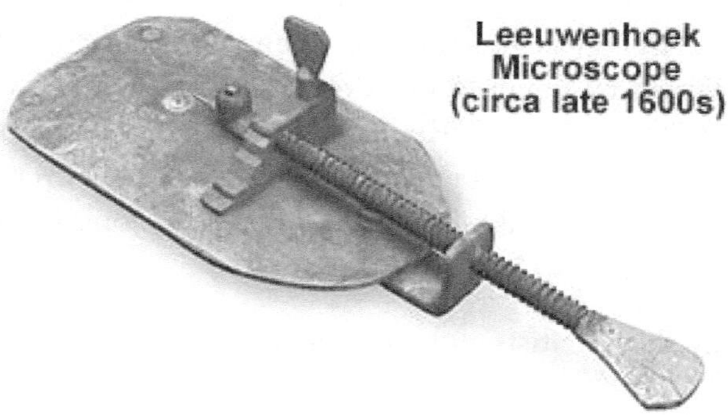

Fig. 2.2: Van Leeuwenhoek's microscope and his description of bacteria

Disprove of Spontaneous Generation Theory

At that time, the age old idea of "Spontaneous Generation theory" was the dominant one. The idea that organism orginate directly from non-living matter. (Life from non-living) also called as abiogenesis (a – not; bio – life; genesis – origin).

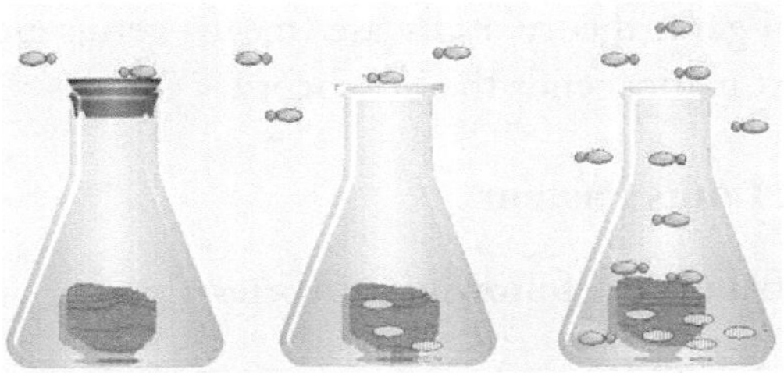

Fig. 2.3: Maggots were developed spontaneously via recombination of matters in rotting materials. (ex meat) (Source: https://www2.nau.edu/gaud/bio301/content/spngen.htm)

Francesco Redi (1626 – 1697): He conducted an experiment to disprove the SG theory. He placed the meat in a jar and covered it with netting and sealed it. The flies laid the egg on the cover/net, and maggots developed on the cover. So He established that the origin of maggots was from flies only, not from meat.

Lazaro Spallanzani (1729 – 1799): He boiled the beef broth for an hour and sealed it in a flask. No microbes observed during incubation.

Franz Schulze (1815 – 1873): He passed the air through a red-hot tube before contact with beef broth leads to no microbial growth during incubation.

Theodor Schwann (1810 – 1882): He passed the air through a red-hot tube before contact with beef broth leads to no microbial growth during incubation.

H. Schroder & T. Von Dusch (1850): He used a cotton plug to filter the air at the mouth of the flask and observed no growth. (Cotton plug for microbial culture and media preparation technique was initiated from now)

Louis Pasteur (1822 – 1895): He performed the "gooseneck experiment." The nutrient of the flask was heated, and the untreated, unfiltered air could pass in or out, but the germs settled in the gooseneck, and no microbes were observed in the nutrient solution.

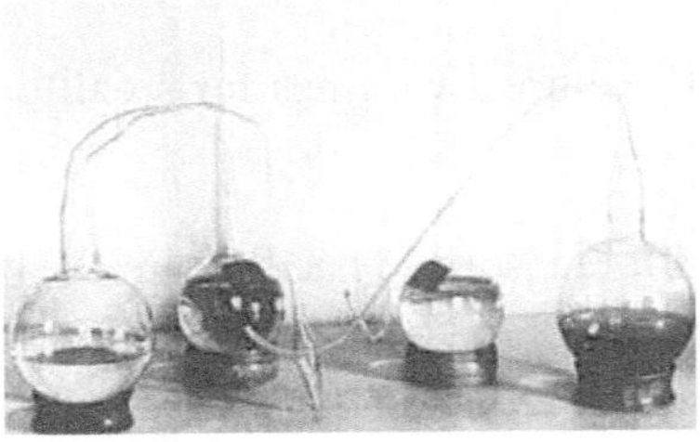

Fig. 2.4: Gooseneck experimental of Louis Pasteur

The concept of germ theory of disease (means germs are responsible for the disease, not the inert matter) ends the SG theory.

Contributions of Louis Pasteur

Louis Pasteur contribute the followings theories-

- Disproved the SG theory
- Discovered that fermenting fruit to alcohol by microbes from that fermentation started
- Sorted different microbes giving different tastes of wine.
- Selected a particular strain (yeast) for high-quality wine.
- Developed a method to remove the undesired microbes from juice without affecting its quality. Heating the juice at 62.80 C for half an hour did the job. This technique is called pasteurisation, which is commonly used in the field of the milk industry.
- Discovered that parasites (protozoa) are causing pebrine disease of silk worms. He suggested that disease-free caterpillars can eliminate the disease
- Isolated the anthrax-causing bacilli from the blood of cattle, sheep, and human beings.
- Demonstrated the virulence (ability of microbe to cause disease) of bacteria
- Developed vaccine (a killed or attenuated microbe to induce the immunity) against rabies from the brains and spinal cord of rabbit.

John Tyndall (1820 – 1893): He proved that dust carries the germs and if no dust in the air, the sterile broth remained free of microbial growth for indefinite period

He also developed a sterilization method "Tyndallization", referred as intermittent or fractional sterilization. The subsequent cooling and heating by steam for 3 days will remove the germs and their spores.

Edward Jenner (1898): Developed vaccines for small pox disease. He coined the term vaccines.

Robert Koch (1843 – 1810): He isolated bacilli from the blood of cattle and infected them with healthy ones and observed the disease. He proved that microorganisms cause the disease. He formulated a sequence procedure to prove that a specific microorganism causes the specific disease, called Koch's Postulate.

Koch's postulates	
1	The agent must be present in every case of the disease.
2	The agent must be isolated and cultured in vitro.
3	The disease must be reproduced when a pure culture of the agent is inoculated into a susceptible host.
4	The agent must be recoverable from the experimentally infected host.

He isolated tubercle bacillus (*Mycobacterium tuberculosis*). He developed a special staining to view Mycobacterium called acid-fast staining. He also discovered and isolated the cholera-causing agent, Vibrio cholerae.

Walther Hesse & his wife Fannie E. Hesse (1883): They used agar instead of gelatin for preparation of media. Agar goes to solution at 100^{0}C and solidifies at 45^{0}C. Till now this was not replaced by any other substance.

Hans Chrisian Gram (1853-1933): He developed a staining procedure for differentiating two different group of bacteria based on the cell wall structure (Gram + ve and Gram – ve). The staining is called Gram staining.

Edvin Klebs & Frederich Loeffler: discovered the diphtheria bacilli and demonstrated the toxin production in the culture flask.

Emil Von Behring & S. Kitasato: Devised to produce immunity by injecting the posion (toxin) into animals so that an antitoxin will be developed.

Elic Metchnikoff: demonstrated that the white blood cells can eat the disease causing bacteria in blood. He called the cells as phagocytes (eating cells) and the phenomenon as phagocytosis.

Paul Erhlich: explained the immunity of human. He also the first used a chemical (arsenic) to control the microbial growth (Chemotherapeutic agent).

Joseph Lister (1878): He developed pure culture technique. Pure culture refers to the growth of a mass of cells of the same species in a vessel. He developed the pure cultures of bacteria using the serial dilution technique.

He also discovered that carbolic acid to disinfect the surgical equipment and dressings leads to the reduction of post-operational deaths/infections.

Alexander Fleming (1928): He identified Penicillium notatum inhibiting Staphylococcus aureus and identified the antibiotic penicillin.

Domagh, (1935): He used sulfanilamide for chemotherapeutic treatment.

Selman A Waksman, (1945): He identified streptomycin antibiotic from soil bacterium. He also coined the term antibiotics (referring to a chemical substance of microbial origin that is in small quantity and exerts antimicrobial activity).

Rene Dubos (1939): He isolated gramicidin and tyrocidin from Bacillus brevis.

Virus Research

» Unknowingly, *Edwart Jenner* handled the small pox virus and controlled by vaccines.

» Same time, L. Pasteur also handled the rabies virus. He also noticed that these microbes cannot be grown in lab condition and can not be seen under microscopes.

» *Charles Chambeland*: developed the porcelain filter, which filter the bacteria but allow the rabies causative agent.

» In 1892, *Iwanowski* discovered the causative agent of tobacco mosaic disease, which was filterable. By spraying the bacteria free filtrate to healthy plant led to tobacco mosaic disease.

» 1935, *Wendell Stanley* crystallized the TMV. These crystals can cause the disease.

» 1936, *F.C. Bawden* explained that the crystalline powder contains protein and nucleic acid.

» *A. Gierer* and *G. Schramm* (1956) proved that the nucleic acid is responsible for infections.

» F.W. Twart and F.d'Herille (1915) discovered the bacteriophages (bacterial viruses).

Soil Microbiology

Helriegel & Wilfarth: found the symbiotic relation between legumes and bacteria.

Leishman (1858): demonstrated that the nodules in legume were formed by bacteria.

Muntz (1878): found the role of bacteria in nitrification in sewage water system.

Martinus William Beijerinck (1851-1931)

- Developed the enrichment technique to isolate various group of bacteria.
- Isolated sulphur reducing bacteria and sulphur oxidizing bacteria from soil.
- Isolated free-living nitrogen fixing bacterium, Azotobacter from soil.
- Root nodulating bacterium, Rhizobium, Lactobacillus, green algae were identified by him.
- He confirmed the Tobacco mosaic virus causes disease and it incorporated in the host plant to reproduce.

Sergei Winogradsky (1856-1953)

The following are the contributions of Winogradsky to soil microbiology.

- Microorganism involed in N cycle, C cycle, S cycle.
- Nitrification process in soil.
- Chemolithotrophy nutrition of soil bacteria.
- Anaerobic nitrogen fixing Clostridium pasteurianum.

J.G. Lopman and P.E. Brown (1903): explained the ammonification process in soil.

L. Hiltner (1904): coined the term "*rhizosphere*" (refers the region around the root surface) where the microbial activity is very high.

J. Dobereiner (1976): isolated Azospirillum, an associative nitrogen-fixing organism that is commonly present in the grasses. The associative symbiotic nitrogen fixation was started from here. She also isolated many associative nitrogen fixers from cereal roots.

B. Frank (1885): coined the term mycorrhiza (association between fungus and higher plant roots).

Barbara Mosse and J.W. Gerdemann: identified the endomycorrhiza which can able to colonize the agricultural crops.

Chapter - 3

About The Microscope

Microscope is an instrument used to magnify and see the enable microorganisms by humanbeeings.

The Microscopes are of two types as Light and Electron microscope.

Basic Principles of Light Microscope

The light is the primary source on which magnification is performed. The magnification is obtained by a system of optical lenses using light waves. These light microscopes can be further classified into

a. Bright field microscope
b. Dark-field microscope
c. Fluorescence microscope and
d. Phase contrast microscope.

Basic units for microscope

Relative size of the microorganisms and their visibility. Man can see about 0.5 mm-sized objects, whereas the light microscopes can be used to visualise up to 1 mm, and electron microscopes (EM) can be used to view 1 nm objects.

Resolving power: The ability to distinguish two objects close to each other. The ability of a microscope in terms of resolving power is called resolution.

For example, the man has the resolving power of 0.2 mm (meaning that he can distinguish two objects with a distance of 0.2 mm close to each other). If he wants to see beyond the limit of his resolving power, further magnification is necessary.

The magnification can be made by lenses.

$$\textbf{Resolving power} = \frac{\lambda}{n (\sin \theta)}$$

Where, λ is the wave length of light source and $n (\sin \theta)$ [NA] is the numerical aperture.

(The resolving power (RP) should be of SMALLER value for good quality microscope).

The resolving power of a microscope can be improved either by reducing the wave length of light or by increasing the $n (\sin \theta)$ value.

e. Reducing the λ

The picture shows the range of the spectrum of a light. If we use violet light, we can get high resolution (meaning the lowest resolving power), but blue light (approx. 450 nm) is sensitive to human beings; the blue light is used in microscopes. The other lights will be removed from the light source using filters.

f. Increasing NA ($n \sin \theta$)

Numerical aperture measures how much light cone spreads out between the condenser and specimen. More spread of light gives less resolving power, meaning better resolution.

The numerical aperture depends on the objective lens of the microscope. (Objective lens refers to the lens nearest to the specimen.) There are two types of objective lenses available in any compound microscope.

i. **Dry objective lens**: The lens can view the specimen without any fluid. Air is the medium between the objective lens and the specimen.

ii. **Oil-immersion lens:** The lens can view the specimen in the presence of immersion oil. This immersion oil is the medium between the lens and the specimen.

The oil immersion objective lens has more NA than the dry objective lens. Because oil can increase the n value. In a dry objective lens, the refractive index (n) is 1.0, because air is the medium, whereas immersion oil has a refractive index (n) of 1.2 to 1.4. So, the resolving power of oil immersion objective lenses is greater than that of dry objectives.

The sub-stage condenser (located below the specimen stage of the microscope) increased the angle of light cone too.

Theoretical limit of R.P. for light scope is 0.2 micrometers.

The limit of resolution: The limit of resolution refers to the smallest distance by which two objects can be separated and still be distinguishable or visible as two separate objects.

$$\text{The resolving power of compound microscope} = \frac{\lambda}{2NA}$$

Calculation

Resolving power

For example If we use a microscope with green light (wavelength of 550 nm) and 0.2 NA objective lens in dry condition, the resolving power can be calculated as:

$$RP = \frac{0.55}{2 \times 0.2} = 1.375 \text{ mm}$$

(Meaning that in this position, the two objects with a distance of 1.375 mm can be distinguishable).

Magnification

The magnification refers the number of times a specimen appeared to be larger than its original size.

For compound microscope, the magnification can be calculated as follows:

Magnification = objective lens magnification x eye piece magnification

If we use a 40x objective lens and a 10x eyepiece for viewing a specimen, then the magnification is 400x. (refers to the fact that the specimen is magnified by 400 times its original size).

Numerical aperture of a dry objective lens when viewing a watery specimen with coverslip. (A) and with water immersion. (B) (If we use oil, the NA will be increased further).

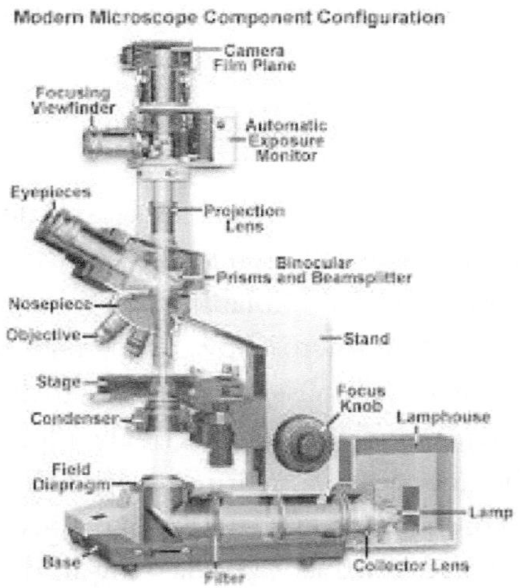

Fig 3.1: The Light Microscope: optical diagram

Parts of typical compound microscope

The optical components are those of all compound microscopes, and even though this model (right side) has no integral lamp, this discussion will begin with the lamp, and proceed in sequence via the mirror to the eye.

The microscope lamp

A microscope requires, especially at the highest powers, intense illumination. The intensity of a light source depends not so much upon its absolute power as upon the amount of light emitted from a given area of the source—lumens per square millimetre rather than just lumens. To achieve high intensity, various lamp designs have come and gone, but one of the commonest and most satisfactory is the low-voltage tungsten filament lamp, with the filament in the form of a tightly-wound, flattened grid. The combination of a suitable quartz-halogen bulb with a concave spherical reflector works well and is also in wide use.

The mirror

The mirror is used only to fold the optical path of the microscope into a convenient space. It also introduces another surface to collect dust. Having said this, the mirror does not need to be of the highest optical quality to make much difference to the quality of the image. A slight film of fine dust (such as remains after dusting the mirror with a blower brush) can actually be useful in locating the beam of light from the lamp when setting up the instrument. Always use the flat side of the mirror in combination with a substage condenser.

The sub stage condenser

The substage condenser fitted to most microscopes is of a design originated by Ernst Abbe in the late 1800s and is usually referred to as the Abbe condenser. While condensers of higher correction are available, the Abbe condenser has proven to be quite satisfactory for routine microscopy. A substage condenser of some kind is an absolute requirement for serious—or at least satisfactory—microscopy. The objectives from x20 upwards require the subject to be illuminated evenly over quite a large angle, and neither a concave mirror nor (especially) a flat mirror is capable of achieving this. If no condenser is used with a high-power objective, the result is an image that is dark, coarse, contrasty, and lacking in detail, described by earlier microscopists as "a rotten image." An important point to note here is that the substage condenser diaphragm is used to control the solid angle of the light emerging from the condenser, illuminating the specimen and filling the objective, not for adjusting the brightness of the image.

The specimen

The specimen is usually supported by a slide and, essentially at the higher powers of the biological microscope, covered by a cover glass. Thus introduced into the image-forming light path, the slide and cover glass become part of the optical system. The thickness of the slide is important to the correction of the substage condenser, and the thickness of the cover glass is critical to the performance of the objective, especially those of higher power (20X and greater). Most substage condensers are corrected to work with a slide thickness of 1.0 mm, and most microscope objectives of x20 or greater power are designed to work with a cover glass thickness of 0.17 mm.

The objective

The objective is the most important component of the optical system in terms of the quality of the final image. The modern microscope objective probably represents the highest degree of optical perfection and precision engineering, which is manufactured in volume for public consumption.

The screw thread of microscope objectives has been a standard across the industry since 1858, when it was first proposed by the Royal Microscopical Society.

The eyepiece

The eyepiece relays to the eye an image projected by the objective into the plane of the eyepiece diaphragm, further magnifying it in the process. In order microscopes, the eyepiece also corrected residual colour errors remaining in the objective. Modern infinity-tube length objectives are fully corrected in themselves but still require additional focusing optics and appropriate eyepieces to produce their image. All of the apochromatic objectives of earlier periods, and many of the higher power achromats, required compensating eyepieces. The other variety of eyepiece in common use is the Huyghenian – best suited to achromatic objectives in general, and particularly to low power achromats which often require little or no correction. These eyepieces are distinguished by a blue fringe around their field diaphragm when the eyepiece is used in bright field.

The eye

The cornea and the eye lens are the final optical components in the image-forming path to the retina. In a person with normal vision, the eye lens will be relaxed as though the eye is forming an image of a very distant object, and the focusing controls on the microscope are used to achieve image sharpness. The optics of the eyepiece used to achieve image sharpness. The optics of the eyepiece are such that all image-forming rays pass through a circle (called the Ramsden disc) a few millimetres exterior to the eyepiece lens and just smaller than the diameter of the pupil. The eye is brought close enough to the eyepiece for the Ramsden disc and the pupil to coincide, at which point the full circular field of the microscope is seen. Learning to hold the head still in this optimum position, especially with a binocular instrument, is one of many skills acquired by the microscopist.

Advantages: convenient, relatively inexpensive and widely available.

Disadvantages: resolving power 0.2 micrometers at best, can recognize cells but not fine details, Needs contrast, cells are mainly watery and don't contrast with their medium.

Easiest way to view cells is to fix and stain

Optical Instrument	**Resolving Power**	**RP in Angstroms**
Human eye	0.2 millimeters (mm)	2,000,000 A
Light microscope	0.20 micrometers (μm)	2000 A
Scanning electron microscope (SEM)	5-10 nanometers (nm)	50-100 A
Transmission electron microscope (TEM)	0.5 nanometers (nm)	5 A

Types of microscopy

The bright field microscope is best known to students and is most likely to be found in a classroom. Visible light is focused through a specimen by a condenser lens and then is passed through two more lenses placed at both ends of a light-tight tube. The latter two lenses each magnify the image. Limitations to what can be seen in bright field microscopy are not so much related to magnification as they are to resolution, illumination, and contrast. Resolution can be improved using oil immersion lenses; lighting and contrast can be dramatically improved using modifications such as dark field, phase contrast, and differential interference contrast. Fluorescence and confocal microscopes are specialised instruments used for research, clinical, and industrial applications.

Other than the compound microscope, a simpler instrument for low magnification use may also be found in the laboratory. This is the stereo microscope, or dissecting microscope. Stereo microscopes usually have a binocular eyepiece tube, a long working distance, and a range of magnifications, typically from 5X to 35 or 40X.

Bright field microscopy

With a conventional bright field microscope, light from an incandescent source is aimed towards a lens beneath the stage called the condenser, through the specimen, through an objective lens, and to the eye through a second magnifying lens, the ocular or eyepiece. Some microscopes have a built-in illuminator, while others use a mirror to reflect light from an external source. The condenser is used to focus light on the specimen through an opening in the stage. After passing through the specimen, the light is displayed to the eye with an apparent field that is much larger than the area illuminated. The magnification of the image is simply the objective lens magnification (usually stamped on the lens body) times the ocular magnification.

Students are usually aware of the use of the coarse and fine focus knobs, used to sharpen the image of the specimen. They are frequently unaware of adjustments to the condenser that can affect resolution and contrast. Some condensers are fixed in position, and others are focusable so that the quality of light can be adjusted. Usually the best position for a focusable condenser is as close to the stage as possible. The bright field condenser usually contains an aperture diaphragm, a device that controls the diameter of the light beam coming up through the condenser, so that when the diaphragm is stopped down (nearly closed), the light comes straight up through the centre of the condenser lens and contrast is high. When the diaphragm is wide open, the image is brighter and contrast is low.

Microscopy with Oil Immersion

Principle

When light passes from a material of one refractive index to material of another, as from glass to air or from air to glass, it bends. Light of different wavelengths bends at different angles so that as objects are magnified, the images become less and less distinct. This loss of resolution becomes very apparent at magnifications of above 400X.

Placing a drop of oil with the same refractive index as glass between the coverslip and objective lens eliminates two refractive surfaces and considerably enhances resolution so that magnifications of 1000X or greater can be achieved.

To use an oil immersion lens, first focus on the area of the specimen to be observed with the high dry (400X) lens. Place a drop of immersion oil on the coverslip over that area and very carefully swing the oil immersion lens and place it over the oil drop. Focus carefully, preferably by observing the lens itself while bringing it as close to the cover slip as possible, then focusing by moving the lens away from the specimen. When in focus, the lens nearly touches the coverslip. The focal plane is so narrow that it is very easy to focus right past it. If you are focusing toward the specimen, you can drive the lens right into it.

When to use oil immersion lenses

Use an oil immersion lens when you have a fixed specimen that is no thicker than a few micrometers. Even then, use it only when the structures you wish to view are quite small—one or two micrometres in dimension. Oil immersion is essential for viewing individual bacteria. It is nearly impossible to view living and motile cells at a magnification of 1000X, except for the very smallest and slowest.

A disadvantage of oil immersion viewing is that the oil must stay in contact, and oil is viscous. A wet mount must be very secure to use oil. Oil immersion lenses are used only with oil, and oil can't be used with dry lenses, such as the 400X lens. Lenses of high magnification must be brought very close to the specimen to focus, and the focal plane is very shallow, so focusing can be difficult. Oil distorts images seen with dry lenses, so once you place oil on a slide, it must be cleaned off thoroughly before using the high dry lens again. Oil on non-oil lenses will distort viewing and possibly damage the coatings.

Phase-Contrast and Dark Field Microscopy

A main obstacle in the microscopy of biological objects is their poor contrast. Only where a contrast exists or where it can be achieved by contrast-enhancing days can structures be made visible. Light-absorbing parts of a preparation weaken the amplitude of the light waves that pass through them. It is thus also spoken of as amplitude preparations. The change of stronger and feebler light is perceived by the eye as a difference in brightness. The invisible parts of the preparation were gone through by the light without a change of amplitude, but the phase of the light may be changed depending on the consistency of the material. This change is due to the altered speed of the lightwaves. Differences in phase can be perceived neither by the eye nor by a photographic film.

The Dutch physicist F. Zernike succeeded in 1935 in converting phase to amplitude differences. He was awarded the Nobel Prize for this achievement in 1953. Today his method is known as phase-contrast microscopy, and it is by now an integral part of nearly all research and many teaching microscopes. Its eminen advantage is that it allows the examination of living objects and thus to follow the processes within cells. It was the phase-contrast microscope that made it possible to make mitosis visible and even to film it.

For the procedure itself, a special condenser with a ring-shaped mask and an additional "phase ring" that is fixed within the back focal plane of the objective is needed. The "phase ring" has two important tasks:

1. It has to achieve an alignment of the brightness of refracted and unrefracted light as the rays that pass through the preparation are weakened in their intensity. In contrast to a conventional image gained with the help of a light microscope, the background of a phase-contrast image is thus dark.
2. The phase difference of most biological preparations is one-quarter of a wavelength or less (lambda/4). The phase ring is constructed so as to achieve an additional difference of lambda/4. The complete phase difference is thus lambda/2, and the crest and trough of refracted and unrefracted rays extinguish each other. A disadvantage of this method is the appearance of light halos around some objects ("halo effect").

Dark Field Microscopy

This method uses a special condenser with a big aperture through which a ligh beam passes through and also the objective. Only if the object is brought into the centre of the light is the light is diffracted, collected by the objective, and used for image formation. Shiny structures are seen in front of a dark background. This method has no very important role in biology but is impressive with crystals.

Fluorescence microscopy

Fluorescence microscopy is based on the fact that some molecules emit part of the light absorbed by them as longer waves. A well-known example is the red fluorescence of chlorophyll. Since the Austrian teacher M. Haitinger made his examinations in the thirties of this 20th century, it is known that a number of so-called fluorochromes exist with which microscopic preparations can be stained so that they emit fluorescence indirectly. In the following decades, a whole range of so-called vital dyes were detected or developed that, when used in low concentrations, could mark specific parts of living cells or of tissues. They enabled researchers to follow the solute transport within cells and tissues or to ascertain the pH of special compartments, like, for example, the vacuole.

There are two ways in which fluorescence microscopes can be constructed: as epifluorescence and as transmission microscopes. The second is the older way of construction. Three components are needed:

1. A strong source of light that emits mainly short lightwaves. Mercury high-pressure lights have proven useful.
2. The first barrier filter: This filter helps to shut off all radiation other than the one that activates the specific dye. It is placed behind the light source within the light cone. Besides, it is advantageous to work with a dark field condenser.
3. Second barrier filter: This filter is brought into the light cone between the objective and eyepiece. It lets through only long wavelengths, which are caused by the emission of the preparation (so-called "secondary radiation").

In the last years, the epifluorescence microscope has replaced the transmission fluorescence microscope more and more. But the transmission fluorescence microscope is still better suited to only weakly magnifying objectives (2.5X, 6.3X). With the epifluorescence, the objective is also a condenser, and the stronger it is, the more intensive radiation can be used. The heart of epifluorescence is a construction within the light cone between the objective and eyepiece that serves to feed activating radiation into the system and is constructed of a first barrier filter, a beam-splitting mirror, and a second barrier filter.

Excitation and fluorescence with chromatic beam splitters. Similar to the interference filters, these are specially coated mirrors used under 450 to the illuminating beam. They reflect certain spectral ranges, while others are completely transmitted. The separating line between reflection and transmission may be set at any point of the spectrum.

1. Exciting radiation
2. Fluorescence emission

Electron microscopy

In 1924, L. de Broglie discovered the wave character of electron rays, thus giving the prerequisite for the construction of the electron microscope. The prototype was built by M. Knoll and E. Ruska. One of the first biological objects depicted was the Tobacco Mosaic Virus (TMV). The first picture of a cell was published in 1945 by K.R. Porter, A. Claude, and E.F. Fullam.

Transmission Electron Microscope (TEM)

The conventional electron microscopy is nowadays called Transmission Electron Microscopy (TEM). We will therefore start with its construction. The ray of electrons is produced by a pin-shaped cathode heated up by current. The electrons are vacuumed up by a high voltage at the anode. The acceleration voltage is between 50 and 150 kV. The higher it is, the shorter the electron waves are, and the higher the power of resolution. But this factor is hardly ever limiting. The power of resolution of electron microscopy is usually restrained by the quality of the lens systems and especially by the technique with which the preparation has been achieved. Modern gadgets have powers of resolution that range from 0.2-0.3 nm. The useful resolution is therefore around 300,000X.

The accelerated ray of electrons passes a drill hole at the bottom of the anode. Its following way is analogous to that of a ray of light in a light microscope. The lens systems consist of electronic coils generating an electromagnetic field. The ray is first focused by a condenser. It then passes through the object, where it is partially deflected. The degree of deflection depends on the electron density of the object. The greater the mass of the atoms, the greater is the degree of deflection. Biological objects have only weak contrasts since they consist mainly of atoms with low atomic numbers (C, H, N, O). Consequently, it is necessary to treat the preparations with special contrast-enhancing chemicals (heavy metals) to get at least some contrast. Additionally, they are not to be thicker than 100 nm, because the temperature is rising due to electron absorption. This again can lead to the destruction of the preparation. It is generally impossible to examine living objects.

After passing the object, the scattered electrons are collected by an objective. Thereby an image is formed that is subsequently enlarged by an additional lens system. Thus formed, the image is made visible on a fluorescent screen, or it is documented on photographic material. Photos taken with electron microscopes are always black and white. The degree of darkness corresponds to the electron density of the candled preparation.

Scanning Electron Microscope (SEM)

The path of the electron beam within the scanning electron microscope differs from that of the TEM. The technology used is based on television techniques. The method is suitable for the depiction of preparations with conductive surfaces. Biological objects have thus to be made conductive by coating them with a thin layer of heavy metal. The power of resolution is normally smaller than in transmission electron microscopes, but the depth of focus is several orders of magnitude greater. Scanning electron microscopy is therefore also well-suited for very low magnifications.

The surface of the objects is scanned with the electron beam point by point, whereby secondary electrons are set free. The intensity of this secondary radiation is dependent on the angle of inclination of the object's surface. The secondary electrons are collected by a detector that sits at an angle at the side above the object. The signal is then enhanced electronically. The magnification can be chosen smoothly, and the image appears a little later on a viewing screen. Finally, some outlines of new and further developments as given:

1. **The high voltage electron microscope:**It operates with an accelerating voltage of 700-3000 kV. Its power of resolution is greater; the preparation can be thicker, and the strain on the preparation is smaller. But the enormous technical expenditure is disadvantageous. Only a few gadgets exist. New results concerning botany have not been gained.
2. **The scanning transmission electron microscope:** In this development of the SEM, the electrons pass through the preparation, and the secondary radiation thus generated is used for image formation. Here the expenditure is large, but it is still worthwhile since the large molecules of nucleic acids or proteins or molecular complexes like viruses can be depicted much better and gentler than with the TEM.

The interpretation of images gained with electron microscopy is increasingly done with computerized interpretation programs. But they are usually only suitable for the reconstruction of regularly recurring patterns and are found more often on a molecular level than on a cellular one.

Chapter - 4

Microorganism Groups

Microscope is an instrument used to magnify and see the enable microorganisms by human beings.

Place of Microorganisms in Living World

Apart from the plant and animal kingdoms, the third kingdom, "Protista, was developed by E.H. Haeckel (1866). The protista includes unicellular organisms that are typically neither plant nor animal. These organisms, the protists, include bacteria, algae, fungi, and protozoa. Bacteria are referred to as lower protists, and others are called higher protists.

R.H. Whittaker (1969) proposed the concept of five kingdoms. He divided the living beings as :

1. Planta (Photosynthetic)
2. Fungi (Nutrient absorption)
3. Animalia (Nutrient ingestion)
4. Protista (Eukaryotic, unicellular) includes microalgae and protozoa
5. Monera (Prokaryotic) includes bacteria

Three domains of life

Based on the molecular taxonomy the living organisms were divided into three domains. The Domain refers the highest level of kingdom.

The Prokaryotae are now divided into two domains, the Bacteria and the Archaea, as different from each other as either is from the Eukaryota or eukaryotes.

None of these groups is ancestral to the others, and each shares certain features with the others as well as having unique characteristics of its own.

D_1: Bacteria – It includes prokaryotic organisms

D_2: Archaea – It includes certain prokaryotes found in enormous numbers, often from unusual habitats such as very hot, very salty or very anaerobic

D_3: Eukaryota –It including eukaryotic organisms

Bergey's manual of systematic bacteriology

Bergey's manual is the international standard for bacterial taxonomy. The entire world of bacteria and archaea were divided into two kingdoms. Kingdom I is Archaeota and the Kingdom II is Bacteria. The manual divide the bacteria into five volumes as:

Volume I- It include archaea, cyanobacteria and phototrophs

Volume II- It include the proteobacteria

Volume III – Bacteria with low G+C, G^+ ives

Volume IV- It include high G+C, G^+ ives

Volume V – Planctomycetes, Spirochaetes, Fibrobacteres, Bacteroides and Fusobacteria.

Prokaryotes: The organism lacking true nucleus and other organelles like mitochondria, golgi body, endoplasmic reticulum etc. are referred as Prokaryotes like Bacteria, Archaea.

Eukaryotes: The organism possessing membrane enclosed nucleus and other cell organelles are referred as Eukaryotes like algae, fungi and protozoan.

Characteristic	**Prokaryotes**	**Eukaryotes**
Diagrammatic representation		
Electron micrograph		
Size of cell	Typically 0.2-2.0 mm in diameter	Typically 10-100 mm in diameter
Nucleus	No nuclear membrane or nucleoli (nucleoid)	True nucleus, consisting of nuclear membrane & nucleoli

Membrane-enclosed organelles	Absent	Present; examples include lysosomes, Golgi complex, endoplasmic reticulum, mitochondria & chloroplasts
Flagella	Consist of two protein building blocks	Complex, consist of multiple microtubules
Glycocalyx	Present as a capsule or slime layer	Present in some cells that lack a cell wall
Cell wall	Usually present; chemically comlex (typical bacterial cell wall includes peptidoglycan)	When present, chemically simple
Plasma membrane	No carbohydrates and generally lacks sterols	Sterols and carbohydrates that serve as receptors present
Cytoplasm	No cytoskeleton or cytoplasmic streaming	Cytoskeleton; cytoplasmic streaming
Ribosome	Smaller size (70S)	Larger size (80S); smaller size (70S) in organelles
Mesosomes	Present	Absent
Chromosome arrangement	Single circular chromosome; lacks histones	Multiple linear chromosomes with histones
Extrachromosomal DNA	Plasmid	Mitochondria and chloroplast
Cell division	Binary fission	Mitosis
Sexual reproduction	No meiosis; transfer of DNA fragments only (conjugation)	Involves meriosis
Site for cellular respiration	Cell membrane	Mitochondria
Locomotion	Rotating flagella and gliding	Undulating flagella and cilia, and also amoeboid movement
Pili	Sex or attachment pili present	Absent

Groups of Microorganisms

The microorganisms are divided into six distinct groups. There descriptions are as

A. Bacteria: The group of phylogenetically related prokaryotic organisms distinc from archaea.

» The size range from 0.5 – 3.0 mm

» They are unicellular

- Presence of cell wall
- Simple internal structures
- They able to grow in artificial media
- They reproduce asexual simple cell division

Importance

- Some are responcible for disease
- They paly important role in nutrient recycling in soil
- They are useful in industrial production of antibiotics, vitamins, dairy products and value added products

B. Archaea: The group of phylogenetically related group of prokaryotic microorganisms which are primitive and differ from bacteria.

- The size ranges from 0.5 – 2.5mm
- They are unicellular
- They have simple internal structures
- They differ from bacteria by absence of peptidoglycon in cell wall
- The most of them live in extreme environment

Importance

- They produce Methane gas
- They used in waste water treatment and energy production
- They produce temperature resistant enzymes for molecular biology and biochemistry

C. Fungi: The group of eukaryotic microorganisms in which chlorophy is absent.

- The size ranges from 5 – 10 mm
- They consist both unicellular as yeast and multicellular organisms as mushroom
- They are saprophytic in nature
- Presence of cell wall
- They have asexual and sexual reproductio

Importance

- They cause plant diseases, animal and human diseases also
- They help in decomposition of organic matter in the soil and help in nutrient cycle
- They cause food spoilage
- The Yeast used for industrial aaplication
- They used in Antibiotic industry
- They used as food like mushroom

D. Protozoa: The group of eukaryotic microorganisms with lack of cell wall.

- The size ranges from 2.0 to 200 mm
- The large numbers microorganisms are survive in soil, fresh water and marine water
- The most are motile
- They reproduce by asexual and sexual processes

Importance

- They play important role in food chain of aquatic environment
- Some of them cause diseases
- They are important in the ecological balance of many microbial communities in soil and aquatic environment
- They used in industrial waste treatment and also in sewage water treatment

E. Algae: The group of eukaryotic microorganisms that contain chlorophy and carry out the photosynthesis.

- They contain different type of pigments
- The size ranges from 1 mm to many meters
- They are unicellular and well as multicellular
- The most occur in aquatic environment
- The reproduction is sexual and asexual process

Importance

- They are food producer of aquatic environment
- They used as food supplements and also in pharmaceutical industry
- They produce economically importance products – agar, alginic acid and carrageenan
- They also cause some disease

F. Viruses: They are non cellular microorganisms.

- The size ranges from 0.015 – 0.2 mm
- Can grow in laboratory condition (needs living host)
- They are obligate parasites in nature
- Lack of metabolic machinery for energy generation and reproductive system for replication
- They depends for energy and reproduction on higher organisms

Importance

- These microorganisms cause diseases in microorganism, human, animal and plants
- The few are used in molecular biology as a vector

Chapter - 5

Bacteria

Bacteria are very small, most being approximately 0.5 to 1.0 mm in diameter. The smallest bacteria are mycoplasmas, as small as 0.2 micrometers. But certain cyanobacteria are quite large; Oscillatoria cells are 7 micrometres in diameter, the size of red blood cells. And certain eukaryotes are very small, only 1 to 2 micrometres.

Epulopiscium fishelsoni, discovered in 1985 in the intestinal tract of sturgeonfish, is an enormous, cigar-shaped cell, as large as 80 x 600 micrometers. Amazingly, this cell is prokaryotic! Initial evidence by EM was hard to believe but confirmed rRNA comparisons with other organisms, a cousin of the Gram-positive Clostridium genus.

Shape

The shape of the bacterium is governed by its rigid cell wall. Typical bacterial cells are spherical, as cocci and coccus; straight rod-shaped, as bacilli and bacillus; and helically curved rods, as spirilli and spirillum. Few bacteria can exhibit a variety of shapes that are referred to as pleomorphic (e.g., Arthrobacter).

Cocci appear in several characteristic arrangements, depending on the plane of cellular and whether the daughter cells stay together or not.

Division along the same plane forms chains; 2 cocci together – Diplococcus.

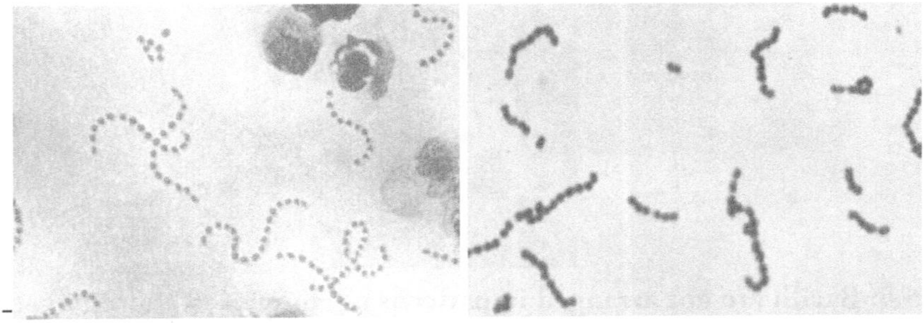

Fig 5.1: 4 – 20 in chains – Streptococcus *spp.* (Source: wikipedia.org)

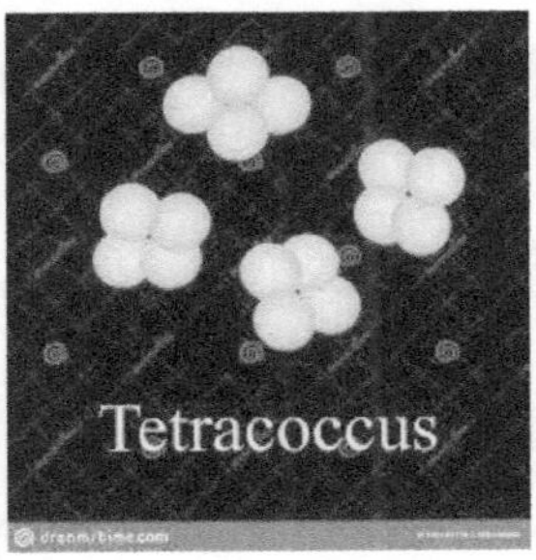

Fig. 5.2: Division along 2 different planes – Tetrads or Tetracoccus (Source: wikipedia.org)

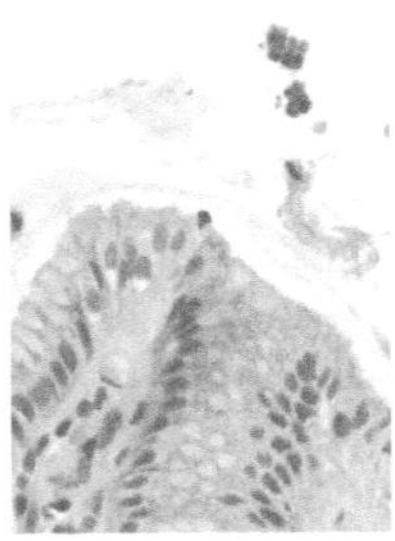

Fig. 5.3: Division along 3 planes regularly – Sarcina (Source: wikipedia.org)

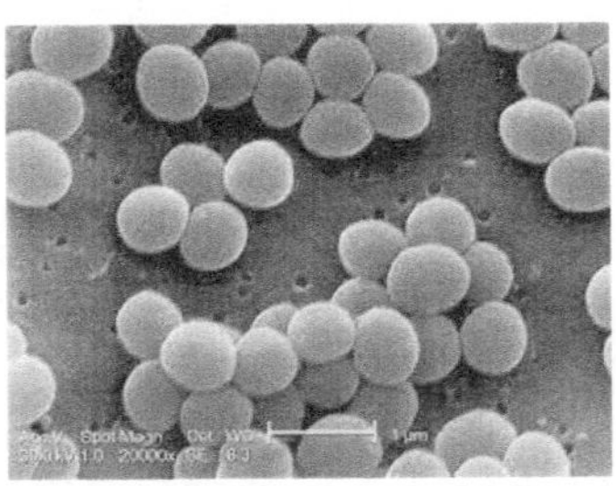

Fig. 5.4: Division along 3 planes irregularly – Staphylococcus (Source: wikipedia.org)

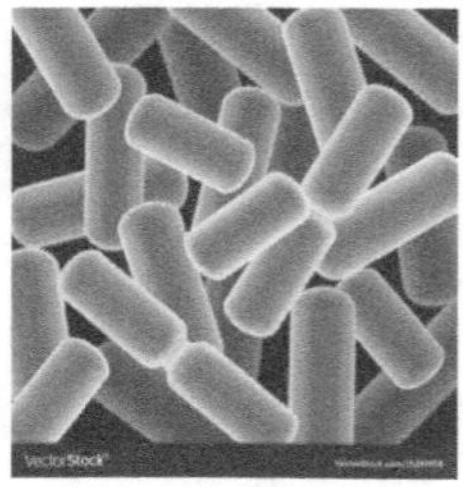

Fig. 5.5: Bacilli are not arranged in patterns as complex as those of cocci and mostly occur or in pairs. The following are some arrangements of rod shaped bacteria. (Source: vectorstock.com)

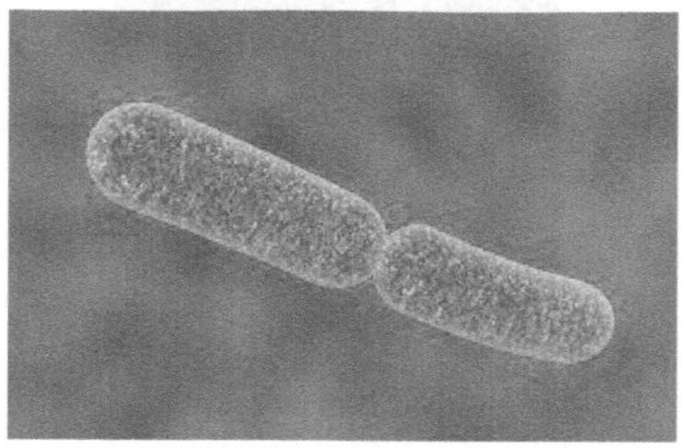

Fig. 5.6: Two bacilli together – Diplobacilli (Source: dreamstime.com)

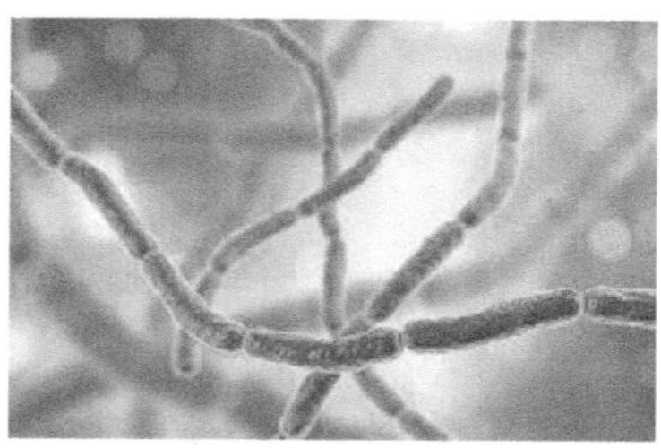

Fig. 5.7: Chains of bacilli are called Streptobacilli (Source: istock.com)

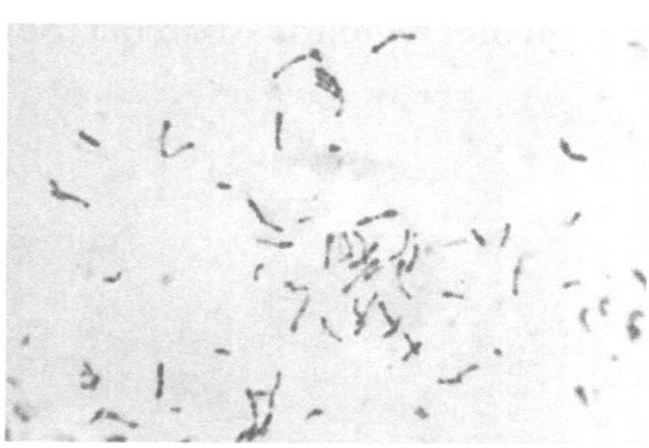

Fig.5.8: Palisades: Rods side by side arrangement like match sticks or in X, V or Y figures. Ex. Corynebacterium diphtheria (Source: wikipedia.com)

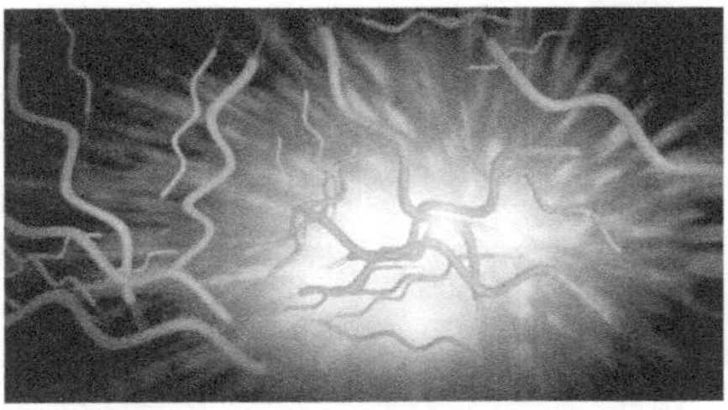

Fig. 5.9: Curved bacteria (spirilla) vary with the number of twists they had. (Source: ScienceDirect.com)

Spiral shape that is rigid is called Spirillium.

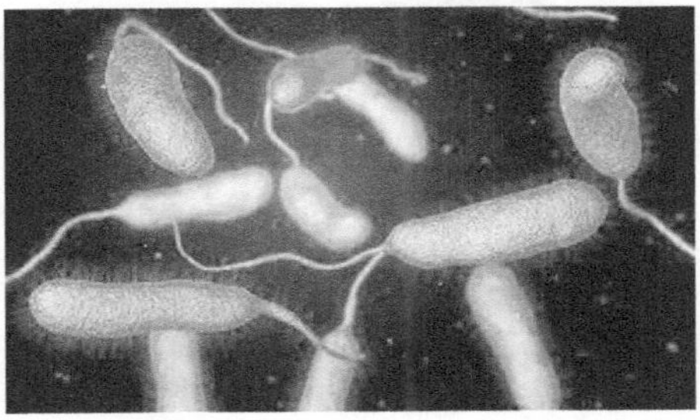

Fig. 5.10: Spirals with less than one complete twist are called as Vibrioid shapes, Ex. Vibrio (Source: food-saftey.com)

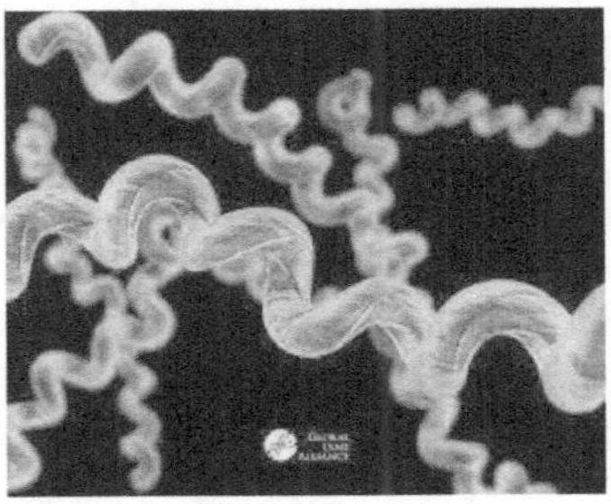

Fig. 5.11: If the organism is flexible and undulating, it is called Spirochete (Source: https://www.globallymealliance.org/blog/unraveling-spirochetes)

Other shapes:

Filamentous shaped: Streptomyces *sp.* form logn, multinucleated, branched filaments called as hyphae

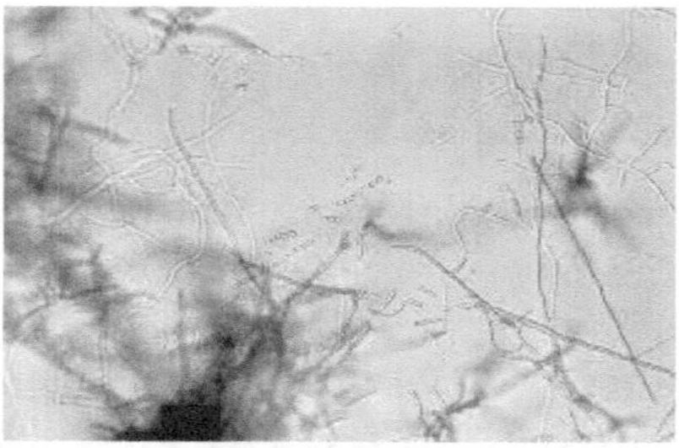

Fig. 5.12: Streptomyces (Source: wikipedia.org)

Pear shaped – Ex. Pasteuria *sp.*

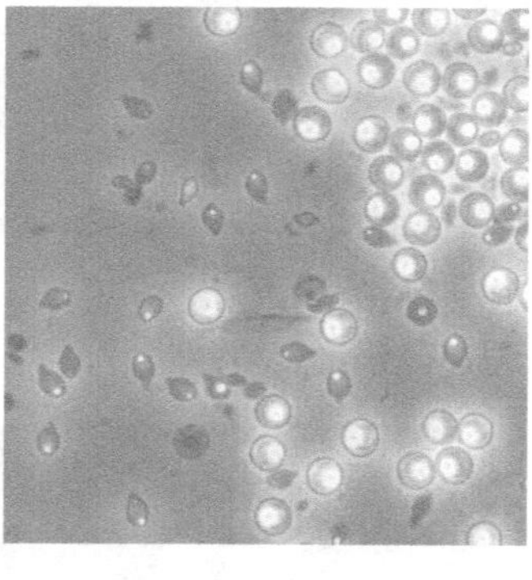

Fig. 5.13: Pasteuria (Source: wikipedia.org)

Lobed spheres – Ex. Sulfolobus *sp.*

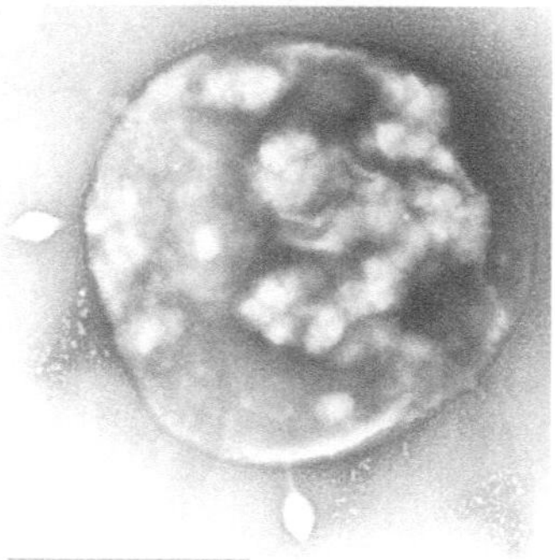

Fig. 5.14: Sulfolobus (Source: wikipedia.org)

Rod with square end – Ex. Bacillus *anthracis*

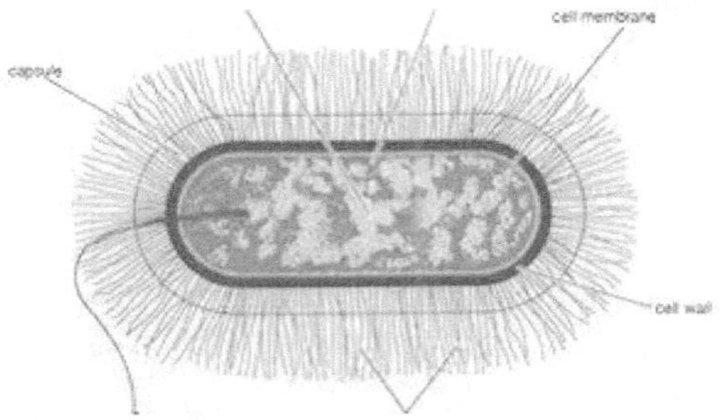

Fig. 5.15: Bacillus *anthracis* (Source: brittanica.com)

Bacteria - Morphology

The Bacteria are unicellular organisms of relatively simple construction. Prokaryotes carry out all cellular functions as individual units.

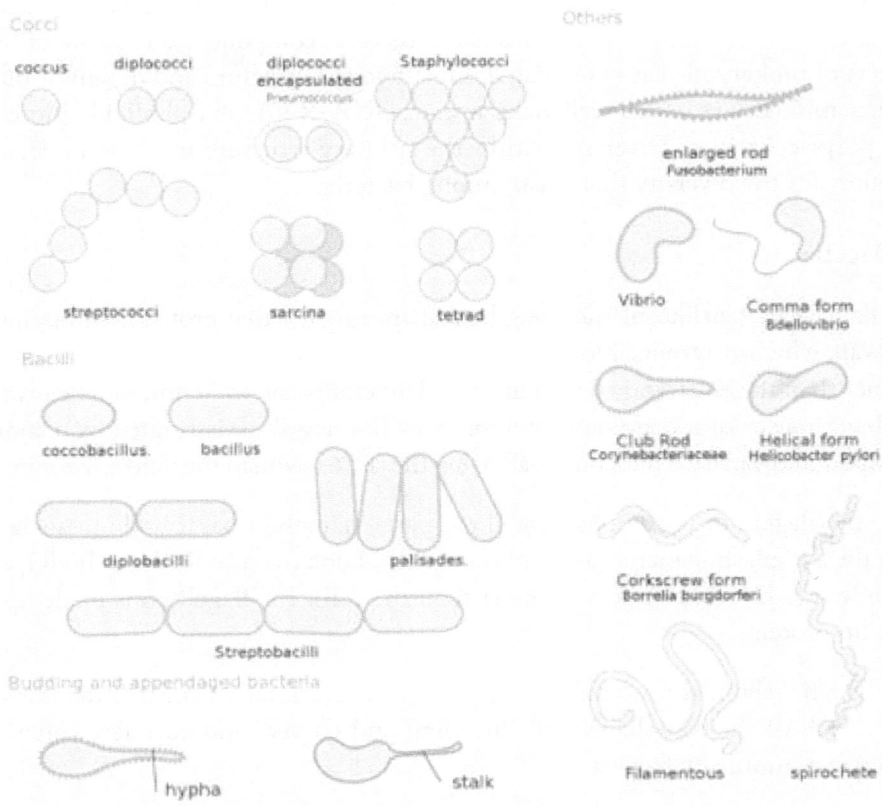

Fig. 5.16: Morphology of Bacteria (Source: wikipedia.org)

A prokaryotic cell has five essential structural components: a genome (DNA), ribosomes, cell membrane, cell wall and a surface layer.

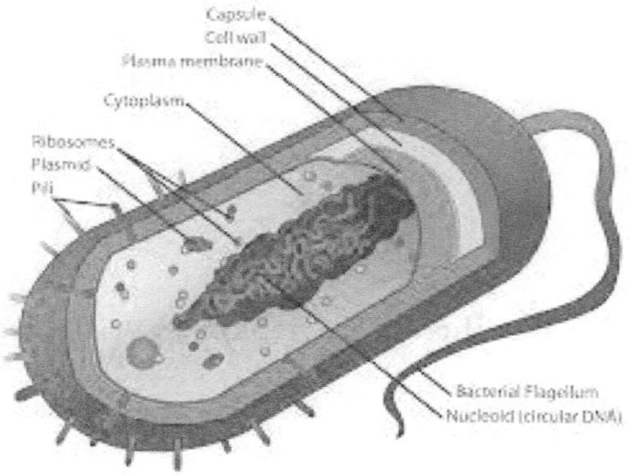

Fig. 5.17: Prokaryotic cell (Source: owlcation.com)

Other than enzymatic reactions, the entire cellular reactions of life can be traced back to the activities of these macromolecular structural components. Thus, functional aspects of prokaryotic cells are related directly to the structure and organisation of the macromolecules in their cell make-up, i.e., DNA, RNA, phospholipids, proteins, and polysaccharides. Diversity within the primary structure of these molecules accounts for the diversity that exists among bacteria.

1. Flagella

The flagella are hairlike, filamentous, helical appendages that protrude through the cell wall, which is responsible for the movement of the bacterium. They are much thinner than the eukaryotic flagellum. The diameter is about 20 nm, well below the resolving power of a compound microscope. The flagellum is rotated by a motor apparatus in the plasma membrane, allowing the cell to swim in the fluid environment.

The flagellum is composed of three parts, namely, a basal body (associated with the cytoplasmic membrane and cell wall), a hook (next to the basal body), and a filament (a hair-like structure longer than the cell). The flagellum is made up of flagellin protein.

The flagellum is powered by the proton motive force of the cell membrane. About half of the rod-shaped, all the spiral and curved, and very few spherical bacteria are motile by flagella.

The distribution of flagellum in the bacteria vary with different types.

Representation of Flagella

a. Monotrichous flagellum: single flagellum present in the pole of the bacterium (Ex. Pseudomonas *aeruginosa*).
b. Lophotrichous flagella: cluster of flagella present in the pole in the bacterium (Ex. Pseudomonas *fluorescens*).
c. Amphitrichous flagella: single or cluster of flagella present in the both pole of bacterium (Ex. Aquaspirillum *serpens*).
d. Peritrichous flagella: flagella surrounded through out the body of the bacterium (Ex. Salmonella *typhi*).

Endoflagella

The flagella, or filaments, present between the outer membrane and inner membrane of the cell wall of the bacterium are called endoflagella, or periplasmic flagella, or axial filaments. They are attached at one end of the cell. Ex. Spirochetes—group of bacteria; Bacterium: Spirochete, Leptospira.

The flagella are responsible for the motility of the bacterium. It rotates like a propeller. Rings in the basal body rotate relative to each other, causing the flagella to turn. The energy to drive the basal body is obtained from the proton motive force. The average speed of bacterial movement is 50 mm/sec, which is equivalent to its 10 body lengths.

2. Pili

Pili are hollow, non-helical, filamentous appendages thinner, shorter, and more numerous than flagella. They are also present in the non-motile bacteria too. Like flagella, they are composed of protein, called pilin. They are also called fimbriae. They are very common in gram-negative bacteria but occur in some archaea and gram-positive bacteria too.

Pili are mainly involved in the adherence or attachment of bacteria to surface, substrates and other cells in nature.

Some pili (F or sex pili) are involved in the transfer of DNA from one cell to other, called conjugation. In E. coli, both sex pili and common pili are present).

They also involve the virulence determination character of a bacterium. Ex. Neisseria gonorrhoeae causes disease if pili are present.

The pili also give resistance to bacterium from phagocytosis (Ex. Streptococcu pyrogenes)

The number of pili varies from few 1-4 to 100 – 200per bacterium.

3. Capsule

Some bacterial cells are surrounded by viscous substance forming a covering laye or envelope around the cell wall, are called as capsules.

If it is visible under light microscope using special staining, it is called as capsul and if it is so thin, not able to see under light microscope, called as microcapsules.

If it is so abundant that many cells are embedded in a common matrix, ar called as slime.

Note: The capsule or slime or microcapsule often referred as Glycocalyx.

Most of the capsules are polysaccharide nature, specifically homopolysaccharide Ex. Streptococcus mutans capsule is made up of glucan.

Some bacterial capsule made up of polypeptides. Ex. Bacillus anthracis capsul composed of polymers of glutamic acid.

Functions

- Provide protection against temporary drying
- Block the attachment of bacteriophages.
- Antiphagocytic (Ex. Streptococcus pneumonia)
- Provide virulence to the bacteria (Ex. S. pnemoniae causes disease only i capsulated).
- Promote attachment of bacteria (Ex. Streptococcus mutans adheres smoot surface of teeth and causes dental caries).
- Prevent cell aggregation in the suspension by electrical charges.
- Important role in biofilm application

Other appendages/structures outside the cell wall

Sheath

Some species of bacteria, particularly those from fresh water and marine environmen from chains or trichomes that are enclosed by hollow tube are called as Sheath. Ex. Sphaerotilus.

Prosthecae

Some species of bacteria have semirigid extension of cell wall and cell membrane are called as prosthecae. This is to increase the surface area of the cell for nutrient absorption. Ex. Calulobacter.

Stalk

The non-living ribbon or tubular appendage that are excreted by the bacterial cell are referred as stalk. Ex. Gallionella. They aided for attachment of the bacteria to surface.

4. Cell wall

All the bacteria have rigid cell walls. The cell wall is the essential structure that protects the cell from mechanical damage and from osmotic rupture or lysis. Prokaryotes live in a relatively diluted atmosphere (which has lower osmotic pressure than the osmotic pressure of inside the cell), which accumulates high salt concentration inside. The osmotic pressure against the inside of the plasma membrane may be the equivalent of 10–25 atm. Since the plasma membrane is a delicate, plastic structure, it must be restrained by an outside wall made of porous, rigid material that has high tensile strength. The cell wall of bacteria deserves special attention for several reasons.

- They are essential for viability
- They are composed of unique compounds no where else in nature
- They are one of the important site for attack by antibiotics
- They provide adherence and receptor sites for drug or virus
- They cause symptoms of disease in animals
- They provide immunological distinction and variation among strains of bacteria

The bacteria can be divided into two major groups based on their cell wall chemistry, called gram-positive and gram-negative bacteria. The gram-negative cell wall is multilayered and quite complex, whereas the gram-positive cell wall consists of a single type of molecule and is often much thicker.

Peptidoglycan layer

The rigid layer of both gram-positive and gram-negative bacteria is similar in chemical composition, called peptidoglycan (or murein). It consists of N-acetyl glucosamine and N-acetyl muramic acid and a small group of amino acids such as L-alanine, D-alanine, D-glutamic acid, and lysine or diamino pimelic acid.

N-acetyl glucosamine and N-acetyl muramic acid were linked by β 1,4 glucoside linkage and the aminoacids were cross linked.

In G +ve bacteria, 90% of the cell wall consists of peptidoglycan layer and few quantities of teichoic acid usually present. Many bacteria have several layers (about 25) of peptidoglycon. In g –ve bacteria, only 10 per cent of wall is peptidoglycon and the major of the wall consists of outer membrane.

The peptidoglycan is signature of the bacteria can be destroyed by certain agents. Ex. Lysozyme, a protein that break the β 1,4 glucoside linkage of peptidoglycan and weaken the cell wall. Water than enters to the cell and cell swells and eventually burst. Lysozyme is present in the tears, saliva and body fluids.

If proper concentration of solute that does not allow the water to penetrate into the cell, such as sucrose, is added to the medium, the solute concentration will be balanced between inside and outside the cell. Under that condition, the lysozyme still digests the peptidoglycan, but lysis does not occur; instead, protoplast is formed. (Protoplast referred to the cell without a cell wall but somatically protected.) If these sucrose-stabilised cells are placed in water, lysis occurs immediately.

Spheroplast often referred as the cell with some remaining of cell wall.

If peptidoglycan layer is removed in G +ve cell, protoplast will be formed and G –ve bacteria will form spheroplast.

Outer membrane

Besides peptidoglycan layer, G –ve bacteria contain additional wall layer made of lipo-polysaccharide. The layer is called as lipopolysaccharide layer or LPS layer or outer membrane.

The lipids and polysaccharides are intimately linked and formed this layer. The polysaccharids has two portions namely, o-polysaccharide and core polysaccharide.

O-polysaccharide contains galactose, glucose, rhamnose, mannose, and one or more dideoxy sugars like abequose, colitose, paratose, or trylose. These sugars were connected in a four- or five-membered sequence that often branched. The core polysaccharide consists of keto deoxyoctonate, glucose, galactose, and N-acetyl glucosamine.

The lipid portion of the LPS, referred to as lipid A, is not a glycerol lipid, but instead, the fatty acids are connected by ester amine linkage to disaccharides composed of N-acetyl glucosamine. The fatty acids are caproic, lauric, myristic, palmitic, and stearic acids.

A lipid protein is also found on the inner side of the outer membrane of gram negative bacteria. It acts like an anchor between outer membrane and peptidoglycan layer.

Endotoxins

In Salmonella, Shigella, Escherichia, the outer membrane causes toxic and pathogenic symptom to human and mammals referred as endotoxins.

Porins

Protein substance presents in the outer membrane of G –ve bacteria. They act like channel for entrance and exit of hydrophobic low molecular substances.

Periplasm

The space in between peptidoglycan and the outer membrane is referred to as periplasm or periplasmic space. In G +ve bacteria, it refers to the space between peptidoglycan and the cell membrane. This space harbours hydrolytic enzymes, binding proteins, detoxifying enzymes, etc.

5. Cell membrane

The cell membrane is the most dynamic structure in prokaryotes. They act as a selective permeability barrier that regulates the passage of substances in and out of the cell.

Bacterial membrane consists of 40 percent phospholipids and 60 percent protein. The phospholipids form a bilayer, and the proteins are present scatterly (called integral proteins), and some proteins are present on the surface of the layer (called peripheral proteins).

The phospholipids are phosphoglycerides in which straight chain fatty acids are ester linked to glycerol.

Functions of cell membrane:

- » Osmotic or permeability barrier
- » Location of transport systems for specific solutes (nutrients and ions)
- » Energy generating functions, involving respiratory and photosynthetic electron transport systems, establishment of proton motive force, and transmembranous ATP- synthersizing ATPase.
- » Synthesis of membrane lipids (including lipoplyscharide in Gram-negative cells).
- » Synthesis of murein (cell wall peptidoglycan).
- » Assembly and secretion of extra cytoplasmic proteins.
- » Coordination of DNA replication and segregation with septum formation and cell division.
- » Chemotaxis (both motility and sensing functions).

6. Mesosomes

The cytoplasmic membrane invaginations in the form of tubular or vesicle-shaped structures are referred to as mesosomes. If the mesosomes are present in the centre of the cell, they are referred to as central mesosomes, and they are involved in the cell division.

The mesosomes that are present in the peripheral region (aside) of the cell are referred to as peripheral mesosomes, and they are involved in the transport of extracellular enzymes from the cytoplasm to external regions.

7. Cytoplasm

The cytoplasm of a bacterial cell is the place where the functions for cell growth, metabolism, and replication are carried out. The gel-like mat, composed of water, protein, enzymes, nutrients, wastes, and gases, like non-cellular materials and chromosomal DNA, ribosomes, and plasmids, like cellular components, are present.

8. Chromosomal DNA

The bacteria do not contain true membrane-enclosed nuclei and chromosomes.

They have a long, coiled, double-stranded, single-circular structure often called Chr. DNA, nucleoid, bacterial chromosome, or chromatin body.

9. Plasmids

A small, circular, covalently closed, self-replicating, extra chromosomal DNA is also present in many bacteria. They are referred to as plasmids. They are mostly controlling the special characters like pathogenicity, nodulation, etc.

10. Ribosomes

The ribosomes are the granular appearance in the cytoplasm involved in the protein synthesis. The size of prokaryotic ribosomes is of 70 S (sedimentation coefficient, Svedberg unit) and is of two sub units of 50S and 30S. The bacterial ribosomes never bound to any organelles in the cytoplasm.

11. Cytoplasmic inclusions

The cytoplasm of bacteria often contains inclusion granules referred to as cytoplasmic inclusions. Inclusion bodies are usually storage materials of energy, carbon, phosphorus, etc. Some aquatic bacteria possess large-sized gas vacuoles, which will help to float them.

The following table shows the cytoplasmic inclusions and their role in bacterial cells.

Some inclusions in bacterial cells

Cytoplasmic inclusions	Where found	Composition	Function
Glycogen	Many bacteria e.g. E. coli	Polyglucose	Reserve carbon and energy source
Polybetahydroxy butyric acid (PHB)	Many bacteria e.g. Pseudomonas	Polymerized hydroxyl butyrate	Reserve carbon and energy source
Polyphosphate (volutin granules)	Many bacteria e.g. Corynebacterium	Linear or cyclical polymers of PO4	Reserve phosphate; possible a reserve of high energy phosphate

Sulfur	Phototrophic purple and green sulfur bacteria and lithotrophic colorless sulfur bacteria	Elemental sulfur	Reserve of electrons (reducing source) in phototrophs; reserve energy source in lithotrophs
Gas vesicles	Aquatic bacteria especially cyanobacteria	Protein hulls or shells inflated with gases	Buoyancy (floatation) in the vertical water column
Parasporal crystals	Endospore- forming bacilli (genus Bacillus)	Protein	Toxic to certain insects
Magnetosomes	Certain aquatic bacteria	Magnetite (iron oxide Fe_3O_4)	Orienting and migrating along geo- magnetic field lines
Carboxysomes	Autotrophic bacteria	Enzymes for autotrophic CO_2 fixation	Site of CO_2 fixation
Phycobilisomes	Cyanobacteria	Phycobili-proteins	Light-harvesting pigments
Chlorosomes	Green bacteria	Lipid and protein and bacteriochlorophyll	Light-harvesting pigments and antennae

Endospores

Certain species of bacteria produce spores either within the cells or are referred to as endospores. They are unique to bacteria, and Bacillus and Clostridium are common spore-forming bacteria.

The endospores are extremely resistant to desiccation, staining, disinfecting chemicals, radiation, and heat. The endospore is able to survive in boiling water, UV light, and many harmful chemicals. They can survive even 100 years.

First, the DNA fragmented, followed by invagination of the cytoplasmic membrane. Then, the portion of the invagination is separated by complete formation of septum (this stage is called forespore). Then the formation of the cell wall around the forespore, followed by cortex, spore coat, and exosporium synthesis, occurred. Finally, the cell lyses, and the spore is released to the external environment.

The outermost layer is called the exosporium, a thin, delicate covering made of protein. The next layer is spore coat, composed of spore-specific proteins. Below the spore coat is the cortex, composed of cross-linked peptidoglycan. Inside the cortex is a spore protoplast, which has a cell wall (core wall), cell membrane, DNA, ribosome, and little cytoplasm. One chemical substance that is characteristic of endospores, not in vegetative cells, is dipicolinic acid in the core region. They occupy 10% of the total dry weight of the spores and are responsible for heat resistance.

Cyst: Cyst is dormant, thick walled desiccation (drying resistant form of cell. It can be differentiated from vegetative cells and can germinate under suitable condition. The cyst don't have temperature resistance. Ex. Cyst of Azotobacter.

Sporangiospores and Conidiospores: The filamentous bacteria, actinomycetes, produce these kinds of spores. The spores were produced at the tip of the hyphae, and if the spores are formed in a sac-like structure (sporangium), they are called sporangiospores; if not, they are called conidiospores. These spores do not have heat resistance but can survive long periods of drying.

Bacteria – Reproduction

Since bacteria are single-celled, a single cell division is enough to get reproduced into two daughter bacteria. The cell division of bacteria is referred to as transverse binary fission. A single cell is divided into two by forming a transverse septum (cross wall). Binary fission is an asexual reproduction.

The process of binary fission is a follows:

1. The cell became elongated from its original size and approximately twice length.
2. DNA replication takes place and DNA moves towards polar region.
3. Septum or cross cell (septum) occurs from the mesosome region
4. New cell wall formation occurs
5. After completion, the cells were separated out

Other methods of reproduction occur in bacteria:

Budding: A small bud developed at one polar region of the cell, elongated and finally developed to new cell. Ex. Rhodopseudomonas *acidophila*.

Fragmentation: Filamentous bacteria (actinomycetes) fragment into small rods and each rod gives raise to new cells. Ex. Nocardia

Conidiospores and sporangiospores: The filamentous bacteria produce spores in the hyphal tips and released to the environment, referred as conidiospores and sporangiospores. If the spores are produced in sac, referred as sporangiospores and without sac are referred as conidiospores.

Growth

In microbiology, growth refers the increase in the number of cells. The bacterial cell is a synthetic machine that able to duplicate itself. More than 2000 chemical processes were occurred during one cell become two.

Step wise reactions occur during the growth of bacteria are as follows:

- » The marcomolecules such as DNA, RNA, protein, polysaccharides, lipids polymerization.
- » Assembly of macromolecules.
- » Cell organelles like cell wall, membrane, ribosome, etc. formation.

Growth rate: Increase of bacterial cell number per unit time is referred as growth rate

Generation: The interval of formation of daughter cells from parental cell is called as generation.

Generation time: Time required to form two daughter cells from a single cell is called as generation time. This is also called as doubling time. E. coli has the doubling time of 20 min and Rhizobium has 2 h.

Growth curve of bacteria

Assuming a single bacterium has been inoculated to a liquid medium and the bacterium starts multiplying by binary fission, it gives an increase of population. If we plot a graph of the log10 value of the population against time, the graph will be as follows:

This curve is referred to as the growth curve of bacteria. The curve can be divided into 4 phases. In the beginning, the growth (increasing the population) will not occur, leading to a straight line (called the lag phase), then the cells start multiplying, giving a linear line (log or exponential phase), followed by a straight line (called the stationary phase) and a declining linear line (death phase or decline phase).

Lag phase: When a bacterium is introduced into a medium, it will not multiply immediately. After a period of time only the bacterium will start multiply. This period is called as lag phase.

» During the period, the cell increases its size.

» It wills synthesis new protoplasm, enzymes, co-enzymes etc.

» The bacterium will adjust the physiological environment.

» The cells are metabolically active.

» At the end of lag phase, each cell will start reproduction.

Log/exponential phase: After lag phase, the cell start multiply steadily by binary fission at constant rate. The portions of curve are linear and have constant generation or doubling time.

» All the cells are having uniform chemical composition, metabolic activity and physiological characters.

» At this time, the generation time will be of constant.

» The cells will be sensitive to chemical inhibitors.

» This is the ideal stage to study the metabolic activity of the bacterium.

Stationary phase: By this time, the essential nutrients of the medium get depleted, and some toxic substances produced by the organism lead to a slowdown in the cell division. At this time, no net increase or decrease of cells in the medium occurs. This stage is called the stationary phase.

» During this stage, cell division rate and cell death rate is uniform.

» Death due to low nutrients content, pH change, toxic wastes, and reduced oxygen occurs.

» Cells are very small and having low number of ribosomes.

» Some cases, cells do not die and at the same time do not multiply.

» Some bacteria produce antibiotics at this stage as secondary metabolites for competitiveness.

Death phase: In this phase, few cells are alive and few died. The death rate is higher than the growth rate leads to decline the population. This phase is referred as death phase or decline phase.

» The cell lysis occur during this stage

» Some species can die very rapidly and some can persists for months

Note: These growth curve and the terms lag, log, stationary and decline phases are used only for the population or group of bacterial cells not for individual cells.

Synchronous growth of bacteria: All the cells of a bacterial culture in same stage of their growth cycle are referred as synchronous growth.

When bacterial growth, which we discussed earlier as a growth curve, is observed under a microscope, it will have different stages of its reproduction, i.e., cell elongation, septum formation, divided cell, etc. So, the cells and cell divisions will not be uniform. But, in synchronous growth culture, the bacterial cells will be of uniform stage.

The synchronous growth of bacteria will have the growth curve as following graph:

The synchronous growth of bacteria can be achieved by growing the cells at suboptimal temperature (below the optimum temperature). E. coli needs 37^{0}C for its normal growth and if grown at 28^{0}C, the synchronous growth will occur. This type of culture will be useful in molecular biology especially the gene cloning and transformation studies.

Study state of growth of bacteria

Maintaining the bacterial population in a particular stage/phase of the cycle is referred to as the study state of growth. The growth curve that was discussed earlier is meant for the bacterial culture in a constant media referred to as batch culture. In this type of culture, the maintenance of bacteria at particular stage is not possible.

For achieving the steady state growth, the constant population should be maintained in the medium. This can be achieved by adding new medium and removing a few cells at frequent time intervals. This type of culture is referred to as continuous culture.

The bacterium at late log phase is much more interesting for many industrial processes. So, maintaining the bacteria at log phase can be achieved by two systems called chemostat and turbidostat.

Chemostat: Maintaining the population of bacteria at a particular stage using an important nutrient source as growth control is referred to as a "chemostat." Ex. In a medium, if the glucose concentration is maintained by subsequent addition at a particular dilution rate, so that it will be taken up by the organism immediately and at the same time a portion of the culture will be removed subsequently. This condition will maintain the bacterial growth at a particular stage, especially in the log phase. This kind of device is called a chemostat.

Turbidostat: An electrical device will monitor the cell density (turbidity), and at appropriate turbidity, the nutrient will be added (as discussed above—glucose) at a dilution rate to maintain the constant population. If the cell density is higher, the instrument will add more quantity of solution to reduce or dilute the population and vice versa. This kind of device is referred to as a turbidostat.

Both the devices are highly useful in the continuous culture of bacteria.

Nutrition and Physical Conditions for Growth of Bacteria

All the bacteria essentially need energy, electron and carbon for their normal cellular activity, growth and reproduction.

Energy: Based on the source of energy, bacteria can be divided into two groups namely, chemotrophs and phototrophs.

a. Chemotrophs- utilize chemicals for their energy.

b. Phototrophs – utilize sun light for their energy.

Electron: Based on the source of electron, bacteria can be divided into two groups namely lithotrophs and organotrophs.

a. Lithotrophs – utilize inorganic substances as electron source.

b. Organotrophs – utilize organic substance as electron source.

Carbon: Based on the sole source of carbon for their utilization, the bacteria can be divided into two groups namely autotrophs and heterotrophs.

i. Autotrophs – use CO_2 as sole carbon source.

ii. Heterotrophs – use organic carbon as sole source.

» The combinations are as follows:

» Photolithotrophs;

» Chemolithotrophs;

» Photoorganotrops;

» Chemoorganotrophs;

» Photoautotrophs;

- » Photoheterotrophs;
- » Chemoautotrophs;
- » Chemoheterotrophs

Some organisms exhibit different nutritional types like Chemo lithotrophi heterotrophs.

(Chemo – chemical for energy; litho – inorganic for electron donor and heter – organic carbon)

Some organisms utilize inorganic compounds as energy and electron donor are inorganic compounds but utilize organic compounds for carbon source, referre as mixotrophs. Ex. Desulfovibrio desulfuricans.

Physical conditions for growth

The prokaryotes exist in nature under enormous range of physical condition such a O_2 concentration, hydrogen ion concentration (pH), temperature, salt concentratio and water activity. The exclusive limit of life on the planet is set by the prokaryote especially the archaea.

a. **Temperature:** Microorganisms have been found growing in virtually al environments where there is liquid water, regardless of its temperature. In 1966 Professor Thomas D. Brock at Indiana University made the amazing discover in boiling hot springs of Yellowstone National Park that bacteria were no just surviving there, they were growing and flourishing. Boiling temperatur could not inactivate any essential enzyme. Subsequently, prokaryotes hav been detected growing around black smokers and hydrothermal vents in th deep sea at temperatures at least as high as 115^{0}C. Microorganisms have bee found growing at very low temperatures as well. In super cooled solutions o H_2O as low as -20^{0}C, certain organisms can extract water for growth, and man forms of life flourish in the icy waters of the Antarctic, as well as househol refrigerators, near 0^{0}C

Based on the growth of bacteria at different temperatures, there are thre different temperatures are available.

1. Minimum temperature: The temperature below which the organism wi not grow at all.
2. Maximum temperature: above which there won't be any growth of organism

3. Optimum temperature: The temperature in which maximum growth occurs.

The following table shows some of the organisms and their min, max and opt. temperature (^{0}C) requirements;

Bacterium	**Minimum**	**Optimum**	**Maximum**
Listeria *monocytogenes*	1	30-37	45
Vibrio *marinus*	4	15	30
Pseudomonas *maltophilia*	4	35	41
Thiobacillus *novellus*	5	25-30	42
Staphylococcus *aureus*	10	30-37	45
Escherichia *coli*	10	37	45
Clostridium *kluyveri*	19	35	37
Bacillus *flavothermus*	30	60	72
Thermos *aquaticus*	40	70-72	79
Methanococcus *jannaschii*	60	85	90
Sulfobobus *acidocaldarius*	70	75-85	90
Pyrobacterium *brockii*	80	102-105	115

The effect of these temperatures on the growth rate of bacteria is represented as the following graph:

Based on the optimum temperature requirement, the organisms can be classified as:

1. Psychrophiles: The organisms which prefer low temperature (15^{0}C) as optimum temperature are referred as psychrophiles. They have about 0^{0}C and 20^{0}C as their minimum and maximum temperature respectively. Ex. Poloromonas vaculata
2. Psychrotolerant: They are the psychrophiles but can grow even at higher temperature of 20-40^{0}C (as the maximum temperature not optimum).
3. Mesophiles: The organisms which prefer 35^{0}C (room temperature) as optimum temperature are referred as mesophiles. They have about 10^{0}C and 48^{0}C as their minimum and maximum temperature respectively. Ex. Bacillus, Pseudomonas, E. coli
4. Thermotolerants: They are the mesophiles but can grow even at higher temperatures (Ex. 450 to 55^{0}C).

5. Thermophiles: The organisms which prefer high temperature (55-65°C) as optimum temperature are referred as thermophiles. They have 400C and 70°C as their minimum and maximum temperatures respectively. Ex. Bacillus *stearothermophilus.*

6. Hyper thermophiles: The organisms which prefer very high temperatures (80-100°C as their optimum temperature is referred as hyper thermophiles. They have 60°C and 110°C as their minimum and maximum temperature respectively. (Ex. Thermococcus).

Optimum growth temperature of some prokaryotes;

Genus and species	**Optimal growth temp (°C)**
Vibrio *cholera*	18-37
Rhizobium *leguminosarum*	20
Streptomyces *griseus*	25
Pseudomonas *fluorescens*	25-30
Erwinia *amylovora*	27-30
Staphylococcus *aureus*	30-37
Escherichia *coli*	37
Mycobacterium *tuberculosis*	37
Pseudomonas *aeruginosa*	37
Streptococcus *pyogenes*	37
Thermoplasma *acidophilum*	59
Thermos *aquaticus*	70
Bacillus *caldolyticus*	72
Pyrococcus *furiosus*	100

Hyperthermophilic Archaea

Genus	**Min. T (°C)**	**Opt. T (°C)**	**Maxi. T (°C)**	**Optimum pH**
Sulfolobus	55	75-85	87	2-3
Desulfurococcus	60	85	93	6
Methanothermus	60	83	88	6-7
Pyrodictium	82	105	113	6
Methanopyrus	85	100	110	7

b. pH: The pH, or hydrogen ion concentration, $[H^+]$, of natural environments varies from about 0.5 in the most acidic soils to about 10.5 in the most alkaline lakes. Most free-living prokaryotes can grow over a range of 3 pH units, about a thousand-fold change in $[H^+]$. The range of pH over which an organism grows is defined by three cardinal points: the minimum pH, below which the organism cannot grow; the maximum pH, above which the organism cannot grow; and the optimum pH, at which the organism grows best. For most bacteria, there is an orderly increase in growth rate between the minimum and the optimum and a corresponding orderly decrease in growth rate between the optimum and the maximum pH, reflecting the general effect of changing $[H^+]$ on the rates of enzymatic reaction.

Microorganisms that grow at an optimum pH well below neutrality (7.0) are called acidophiles. Those that grow best under alkaline neutrophiles and those that grow best under alkaline conditions are called alkaliphiles. Obligate acidophiles, such as some Thiobacillus species, actually require a low pH for growth since their membranes dissolve and the cells lyse at neutrality. Several genera of Archaea, including Sulfolobus and Thermoplasma, are obligate acidophiles. Among eukaryotes, many fungi are acidophiles, and the champion of growth at low pH is the eukaryotic alga Cyanidium, which can grow at a pH of 0.

The following table shows the mini. pH, opt. pH and maxi. pH requirement of few bacteria.

Organism	**Minimum pH**	**Optimum pH**	**Maximum pH**
Thiobacillus *thiooxidans*	0.5	2.0-2.8	4.0-6.0
Sulfolobus *acidocaldarius*	1.0	2.0-3.0	5.0
Bacillus *acidocaldarius*	2.0	4.0	6.0
Zymomonas *lindneri*	3.5	5.5-6.0	7.5
Lactobacillus *acidophilus*	4.0-4.6	5.8-6.6	6.8
Staphylococcus *aureus*	4.2	7.0-7.5	9.3
Escherichia *coli*	4.4	6.0-7.0	9.0
Clostridium *sporogenes*	5.0-5.8	6.0-7.6	8.5-9.0
Erwinia *caratovora*	5.6	7.1	9.3
Pseudomonas *aeroginosa*	5.6	6.6-7.0	8.0
Thiobacillus *novellus*	5.7	7.0	9.0
Streptococcus *pneumonia*	6.5	7.8	8.3
Nitrobacter *sp.*	6.6	7.6-8.6	10.0

c. **Oxygen:** Oxygen is a universal component of cells and is always provided in large amounts by H2O. However, prokaryotes display a wide range of responses to molecular oxygen O2.

Molecular oxygen is both beneficial and harmful for biologicals.

Beneficial: It act as terminal electron acceptor in the respiratory chain – Electron transport chain reaction enable to form ATP.

Harmful: The oxygen derivatives such as hydrogen peroxide, singlet oxygen and super oxide are highly toxic and oxidizing agents.

The organisms can be divided into 4 or 5 different groups based on their oxygen requirement.

Obligate aerobes require O2 for growth; they use O2 as a final electron acceptor in aerobic respiration. Ex. Micrococcus

Obligate anaerobes: Do not need or use O2 as a nutrient. In fact, O2 is a toxic substance, which either skills or inhil it's their growth. Obligate anaerobic prokaryotes may live by fermentation, anaerobic respiration, bacterial photosynthesis, or the novel process of methanogenesis. Ex. Clostridium, Methanobcaterium.

Facultative anaerobes (for facultative aerobes) are organisms that can switch between aerobic and anaerobic types of metabolism. Under anaerobic conditions (no O2) they grow by fermentation or anaerobic respiration, but in the presence of O2 they switch to aerobic respiration. Ex. E. *coli*.

Aerotolerant anaerobes are bacteria with an exclusively anaerobic (fermentative) type of metabolism but they are insensitive to the presence of O2. They live by fermentation alone whether or not O_2 is present in their environment. Ex. Streptococcus.

Microaerophiles: They need very low level of oxygen for their growth, but cannot tolerate the level of atmospheric O_2 level. Ex. Spirillum *volutans*.

Table 6. Terms used to describe O_2 relations of Microorganisms

	Environment		
Group	**Aerobic**	**Anaerobic**	**O_2 effect**
Obligate Aerobe	Growth	No growth	Required (utilized for aerobic respiration)
Microaerophile	Growth if level not too high	No growth	Required but at level below 0.2 atm
Obligate Anaerobe	No growth	Growth	Toxic
Facultative Anaerobe (Facultative Aerobe)	Growth	Growth	Not required for growth and utilized when available
Aerotolerant Anaerobe	Growth	Growth	Not required and not utilized

Oxygen detoxification: Most of the aerobes (and aerotolerant organisms too) have three major enzymes either all or two to detoxify the oxygen derivatives. They are catalase, peroxidase and super oxide dismutase. Obligate anaerobes lack these enzymes, and therefore undergo lethal oxidations by various oxygen radicals when they are exposed to O2.

d. **Water activity:** Water is the solvent in which the molecules of life are dissolved, and the availability of water is therefore a critical factor that affects the growth of all cells. The availability of water for a cell depends upon its presence in the atmosphere (relative humidity) or its presence in solution or a substance (water activity). The water activity (Aw) of pure H2O is 1.0 (100% water). Water activity is affected by the presence of solutes such as salts or sugars that are dissolved in the water. The higher the solute concentration of a substance, the lower is the water activity and vice-versa. Microorganisms live over a range of Aw from 1.0 to 0.7. The Aw of human blood is 0.99; seawater = 0.98; maple syrup = 0.90; Great Salt Lake = 0.75. Water activities in agricultural soils range between 0.9 and 1.0.

The water requirement (Aw) of few bacteria is as follows:

Caulobacter – 1.00; Pseudomonas, Salmonella, E. coli – 0.91; Lactobacillus – 0.90; Bacillus – 0.90; Staphylococcus – 0.85; Halococcus – 0.75.

Note: The concept of lowering water activity in order to prevent bacterial growth is the basis for preservation of foods by drying (in sunlight or by evaporation) or by a addition of high concentrations of salt or sugar.

e. **Salt concentrations:** The only common solute in nature that occurs over a wid concentration range is salt [NaCI], and microorganisms are named based o their growth response to salt. Microorganisms that required some NaCI fo growth are halophiles.

Mild halophiles require 1-6% salt, (Ex: Staphylococcus areus)

Moderate halophiles require 6-15% salt; (Ex. Vibrio fisheri).

Extreme halophiles require 15-30% NaCI for growth. (Ex. Halobacteriun salanarum).

Bacteria that are able to grow at moderate salt concentrations, even though the grow best in the absence of NaCI, are called halotolerant. Although halophiles ar "osmophiles" (and halotolerant organisms are "osmotolerant") the term osmophile is usually reserved for organisms that are able to live in environments high in suga

Organisms which live in dry environments (made dry by lack of water) ar called xerophiles.

Archaea

Archaea are the diversified group of prokaryotes primitive to bacteria.

The phylogenetic tree of life consists of three domains (the highest unit o taxonomy) namely, a. archaea b. bacteria c. eukaryote. Among the three, archae and bacteria are prokaryotes and eukaryote consists of all eukaryotic organism including plant and animals.

The archaea decide the limit of life on the earth. These organisms survive or the maximum extreme conditions occur on the earth. Ex. High temperature ho springs; high acidic soils; high salt lakes, extreme anaerobic conditions etc.

Phenotypic properties of Bacteria and Archaea compared with Eukarya

Property	Cellular Domain Eukarya	Bacteria	Archaea
Cell configuration	Eukaryotic	Prokaryotic	Prokaryotic
Nuclear membrane	Present	absent	absent
Number of chromosomes >1		1	1

Chromosome topology	linear	Circular	Circular
Murein in cell wall	-	+	-
Cell membrane lipids	Ester-linked glycerides; unbranched; polyunsaturated	Ester-linked glycerides; unbranched; saturated or monounsaturated	Either-linked branched; saturated
Cell membrane sterols	Present	Absent	Absent
Organelles (mitochondria and chloroplasts)	Present	absent	Absent
Ribosome size	80S (cytoplasmic)	70S	70S
Cytoplasmic streaming	+	-	-
Meiosis and mitosis	Present	Absent	Absent
Transcription and translation coupled	-	+	+
Amino acid initiating protein synthesis	Methionine	N-formyl methionine	Methionine
Protein synthesis inhibited by streptomycin and chloramphenicol	-	+	-
Protein synthesis inhibited by diphtheria toxin	+	-	+

Morphological similarities & variations between archaea and bacteria

When observed under microscopes, archaea and bacteria look similar but few morphological similarities and variations occur between archaea and bacteria.

Similarities

» The archaeal cell size, shape and structures and similar to bacteria.

» Both contain single circular DNA as chromosomal material

- Both contain 70s ribosomes.
- Cell well and cell membranes are present in both the cases.

Variations

- The cell wall of archaea is made up of proteins, glycoproteins, and polysaccharides. Few archaea have peptidoglycan layers but with modifications. In bacteria, the N-acetyl glucosamine and N-acetyl muramic acid are linked by means of β 1-4 linkage, whereas in archaea, instead of NAM, N-acetyl talosaminuronic acid is linked with N-acetyl glucosamine with β 1-3 linkage. This type of cell wall is referred to as pseudomurein or pseudopeptidoglycan layer.
- Since, the cell wall composition is different; the lysozyme can not act on the cell wall, whereas the bacteria can be lysed by lysozyme.
- The bacterial cell membrane is made up of long chain fatty acid which has ester linkage with glycerol. Whereas the archeal cell membrane is a long chain alcohol with ether linkage with glycerol.
- Methionine is the first amino acid in archeal protein synthesis whereas the bacterium uses N-formyl methionine as its first aminoacid formation.
- Histone is present in the chromosome material of archaea, whereas the bacteria do not have histone protein.

Diversity

Like bacteria, the archaea also has variety of metabolism. They have chemoorganotrophy, chemolithotrophy type of nutrition. Based on the habitat and phylogenic relationship, the archaea are divided into four groups as:

1. Extreme halophilic archaea
2. Methanogens
3. Hyper thermophiles
4. Thermoplasma

1. Extreme halophilic archaea

These archaea survive in extremely higher salt concentrations like Great salt lake (USA). The salt concentration is about 40.1 g/1 of Na^+ and 181 g/1 of CI. These archaea has the optimum salt concentration of 1.5 M to 5.5 M NaCI. These archaea are collectively called as halobacteria.

The genera are Halobacterium, Halorubrum, Halococcus, Natronobacterium, Natronosomonsa.

- These bacteria are heterotrophs.
- They have aerobic respiration with electron transport chain
- All the organelles were stabilized by K+.
- They provide purple membrane in the cell membrane, which is used for ATP synthesis
- Bacteriorhodopsin is the light-harvesting pigment present in the membrane that gives a purple colour. These pigments directly produce the proton gradient in the membrane (as that of the electron transport chain) but without any electron carriers, finally leading to ATP synthesis. This process is known as non-photosynthetic phosphorylation.

Protection against salt;

- Internal accumulation of K ions
- Acidic aminoacids of cell membrane
- K stabilized cell organelles: are some of the mechanisms by which these archaea protected them from salt.

2. Methanogens

They are the obligate anaerobic thermophiles present in fresh water sediments, deep soil, ruman of cattle, paddy fields. (Cattle can produce 50 l of methane per day).

- They are the obligate anaerobes
- They use H_2 and CO_2 as electron and carbon sources for energy generation and anabolism through acetyl CoA pathway.
- Some organisms use CO, CH_3OH, and HCHO as other substrates for C assimilation.
- They use CO_2 as terminal electron acceptor for methane formation.
- The role of methanogens in decomposition of organic waste is highly useful in environment and at the same time methane emission from flooded soil is highly harmful to ozone layer of atmosphere.

Methanobacterium, Methanobrevibacterium, Methanococcus, Methanospirillum, Methanosarcina are some of the important genera present in this group.

3. Hyperthermophiles

They are extreme thermophiles present in hot sulphur rich environments such as hot springs, volcanisms etc.

- » They require 80-100^{0}C as optimum temperature for growth.
- » Unusually stable at high temperatures.
- » They need sulphur as essential nutrients.
- » Both aerobes and anaerobes are present.
- » Chemolithotrophy and chemoorganotrophy type nutrition.
- » Chemolithotrophs reduce the sulphur as terminal electron acceptor, form H2S and derive energy. Ex. Pyrococcus, Ferroglobus, Archeoglobus.
- » Chemoorganotrophs use organic compounds and sulphur for energy and carbon source. Ex. Thermoproteus, Thermococcus, Desulfurococcus, Pyrococcus, Sulfolobus.

Sulfolobus was the first hyperthermophile isolated by T.D. Brock from hot spring of Yellowstone National Park.

Thermos aquaticus is an archaean from which the DNA polymerase enzyme isolated, is very much useful in the DNA synthesis in lab condition celled as Polymerase chain reaction.

4. Thermoplasma & Picrophilus

They are thermophilic acidophiles. They prefer 55^{0}C as optimum temperature and pH 2.0 as optimum condition for growth. They are present in the coal mine spoils (waste created during coal mining).

Another feature of these archaea is lack of cell wall. They have only membrane and the membrane itself protect the cell from high temperature and acidity.

Bacterial Taxonomy

Taxonomy refers the science of classification, which includes identification, nomenclature and systematic arrangement of any group of organisms.

Identification – Identify the individual

Nomenclature – naming the individual

Classification – arranging the individual by groups

Conventional classification method

For identification, nomenclature and classification of any bacterium, conventionally few characters are used. The characters are as follows:

1. Morphological characters: Cell shapes, size and structures; cell arrangement; occurrence of special structures; developmental forms; staining reactions; motility and flagellar arrangement.
2. Chemical characters: Various chemical constituents of cell like cell wall chemicals.
3. Cultural characters: Nutritional requirement and physical conditions required for growth and the manner by which growth of bacteria occur in solid and liquid medium.
4. Metabolic characters: The way by which the bacterium obtain their energy, carbon and electron source; regulation of these reactions.
5. Antigenic characters: Special large chemical components (antigens) of the bacterium – distinctive for certain kind of microorganisms.
6. Genetic characters: Characteristics of hereditary material of the cell (DNA) and occurrence and functions of other kind of DNA that may be present such as plasmid.
7. Pathogenicity: The ability to cause the disease in various plant and animal or even other microorganisms.
8. Ecological characters: Habitat and distribution of the organism in nature and the interaction between and among species in natural environment.

Based on these above characters, the conventional taxonomy of bacteria developed.

Numerical Taxonomy: A method used in taxonomy to determine and numerically express the degree of similarity of every strain of prokaryotes is referred as numerical taxonomy.

Recent method of bacterial classification

To understand the present concept of classification, the knowledge of evolution is essential. The present bacterial classification is mainly based on the evolutionary relatiohships of the bacteria.

A brief about evolution:

- 4.6 billion years ago, the earth was originated
- 3.86 billion years ago, water formation occur
- 3.5 billion years ago rocks formed
- Serveral shells of earth were formed. 4 layers. 1. Atmosphere- H2O, Nitorge CO2, SO4, CH4 2. Crust – U, K, O, Na, TH; 3. Mantle – Si, AL, O, M Fe; 4. Core: Fe, Si, Ni)
- At that time, (3.5 b years) in the presence of lightening & UV light methane, ammonia, hydrogen, water were converted to biochemicals lik amino acid, formaldehydes.
- Then formed RNA, able to replicate them selves.
- Then surrounded by lipoprotein for the new life. (RNA world)
- Then DNA formed and RNA acts as a coding agent. Protein played majo role and thus a new life formed.
- The atmosphere of that period was reduced one.
- The fossils are the evidence for microbial life of that period. Ancient rock contain recognisable microfossils that appear as bacterial shapes. They ar referred to as stromatolites. These fossils were recognised as filamentou phototrophic bacteria (not as present cyanobacteria—because the presen one has oxygenic photophosphorylation, whereas at that time the atmospher was devoid of oxygen). The single-celled phototrophs formed about 3.. billion years ago.
- About 1 billion year's agro, eukaryotes formed.
- 600 million years ago variety of plants and animals formed, evolved an disappeared.
- Mammals formed by 100 million years ago and 2-3 million years, ago MAN appeared.

Recent method of classification of prokaryotes is mainly based on th evolutionary relationship. There are three methods presently available for grouping the microorganisms.

- DNA: DNA hybridization
- Ribotyping (16s RNA analysis)
- FAME (Fatty acid methyl ester analysis)

Among the above three, ribotyping is most common and frequently used tool to classify the microorganisms.

1. **DNA:** DNA hybridization: The DNA of one organism is subjected to hybridize with other organism and the degree of hybridization will vary with organisms depends upon their relativity. This variation will be used to identify and group the organism.
2. **FAME:** The fatty acid composition of prokaryotes gives very high diversity. The fatty acid compositions especially the cell wall fatty acids analysis is used to identify the organisms.
3. **Ribosomal analysis:** Among the cellular organelles, ribosome is present in all the living organisms; ancient molecule; functionally constant; universally distributed and well conserved.

Ribosomal RNAs in Prokaryotes

Name	Size (nucleotides)	Location
5S	120	Large subunit of ribosome
16S	1500	Small subunit of ribosome
23S	2900	Large subunit of ribosome

The ribosome can be divided into two subunits namely 30S and 70S (In eukaryotes 80S). The 30S subunit is referred as Smaller Sub-Unit (SSU). Further, the 30S can be divided in to 3 units namely, 5S, 16S and 23S. (Based on their sedimentation co-efficient). (Note: For eukaryotes, 18S instead of 16S).

By analyzing the genetic information of 16S RNA, the phylogeny of a prokaryote is possible. Phylogeny refers the evolutionary history of the organism.

The 16S RNA genetic analysis can be done by following procedure:

After getting the gene sequence information, the DNA base information will be compared with already analysed databases. Based on the similarity percentage, the evolutionary distance of an organism will be calculated. Based on the evolutionary distance of various organisms, a branched tree-like map can be obtained, which is referred to as a phylogenetic tree.

The 16S rRNA sequence that is unique to an individual organism at the strain level is often referred to as a signature sequence. By comparing the genetic information of 16S RNA, it is possible to identify and classify the organisms.

Three domains of life

Based on the signature sequence, the phylogenetic tree is divided into three domains, namely, bacteria, archaea, and eukaryota. Among the three, two of which are prokaryotes (bacteria and archaea). This universal phylogenetic tree is sharing a common ancestor called the universal ancestor, which is at present not known. The microorganisms fall in all three domains.

Domain: Eucarya (Eukaryotes)

Major Divisions of Eucarya

The Eucarya are usually classified into four kingdoms: plants, animals, fungi, and protists. The first three of these correspond to phylogenetically coherent groups as well. However, the eukaryotic protists do not form a group but rather are comprised of many phylogenetically disparate groups (including slime moulds, multiple groups of algae, and many distinct groups of protozoa). Just as molecular analyses were required to see the natural relationships among prokaryotes, they are also allowing us to infer relationships among these non-plant, non-animal, non-fungus eukaryotes.

The origin of eukaryotes from prokaryotes was mainly explained by the endosymbiont theory. The chloroplasts and mitochondria of eukaryotes contain smaller amounts of DNA. These 16S rRNA sequence analyses showed that mitochondria and proteobacteria like Agrobacterium and Rhizobium are close relatives; Rhodopseudomonas and Rhodobacter are phylogenetically highly related. This theory was evidenced by 16S RNA sequence and phylogenetic analysis.

Domain: Archaea

Major Groups within the Archae (Note: Please refer the notes on archaea for details).

The major physiological groups within the Archaea (informally, archaea) are extreme thermophiles (including thermoacidophiles), sulphate reducers, methanogens, and extreme halophiles. These physiological groups do not correspond to the phylogenetic groups. This is largely because the methanogens have given rise to multiple other groups, including the halophiles, the sulphate reducers, and at least two other lineages of nonmethanogenic thermophiles.

The thermophilic members of the Archaea include the most thermophilic organisms cultivated in the laboratory, some having optimal growth at >100 EC. The aerobic thermophiles are also acidophilic; they oxidise sulphur in their environment to sulphuric acid, creating acid hot springs. The hot acidic waters can dissolve limestone and some other minerals out of the surrounding rock matrix, creating boiling mud pots of the insoluble residue from the rock.

The extreme halophiles include the most salt-tolerant organisms known. They are aerobic or microaerophilic. They have also invented a form of photosynthesis that does not use chlorophyll.

The sulphate-reducing Archaea are now known to be a widespread group. One of the reasons that this group has received attention is that sulphate-reducing organisms are responsible for "souring" oil wells (increasing their sulphur content). In addition to increasing the sulphur emissions when the oil is burnt (or increasing the cost of refining it), sulphide attacks the metal in well casings and pipelines.

Methanogens are strict anaerobes, yet they gave rise to at least two separate aerobic groups: the halophiles and a thermoacidophilic lineage. Methanogens produce most of the methane in the Earth's atmosphere. Methane is a significant greenhouse gas (i.e., it is a cause of global warming). Much of this methane production is associated with activities for which humans are at least partially responsible. This includes the flooding of rice paddies and the raising of domestic cattle.

Domain: Bacteria

This domain contains the prokaryotic organisms, bacteria. They are a diversified group that varies in morphology, metabolism, nutritional, and physiological conditions.

Nomenclature

Following the binomial system of nomenclature, all the bacteria are given genus and species names. The names are either Latin or Greek derivations of some descriptive property appropriate for the species. Ex. Bacillus subtilis, B. cereus, B. sterothermophilus, and B. acidocaldarius mean "slender," "waxan," "heat-loving," and "acid-thermal," respectively.

Taxonomical groups

Linnaeus defined the biological classification system that we still use for plants and animals and, with relatively minor modifications, for fungi and other microorganisms. It is a hierarchical system that starts with a few categories at the highest level and further subdivides them at each lower level. The levels in the hierarchy are as follows:

Domain							
	Kingdom						
		Phylum					
			Class				
				Order			
					Family		
						Genus	
							Species

Bergey's Manual of systematic bacteriology

Bergey's manual of systematic bacteriology is a compendium of standard and molecular information about the available prokaryotes. The Bergey's manual of systematic bacteriology – First edition consists of 4 volumes:

Vol 1. Gram positive bacteria

Vol 2. Gran negative bacteria

Vol 3. Bacteria with unusual properties including archaea

Vol 4. Filamentous bacteria

This system does not have much phylogenic information. The second edition of the Bergey's manual provides much clear information about the genetic relationship among the organisms. The second edition was divided into 5 volumes.

Vol 1. Archaea, cyanobacteria, phototrophs and deeply branched genera

Vol 2. Proteobacteria

Vol 3. Low G+C gram positives

Vol 4. High G+C gram positives

Vol 5. Planctomycetes, Spirochetes, Bacteroides, Fusobacteria

The examples of each group of bacteria are as follows:

Vol 1. Archaea, cyanobacteria, phototrophs and deeply branched genera

This volume has 3 important groups out of which one is in different domain (Domain – Archaea).

. Archaea

Four important sections are present in the archaea.

a. Hyperthermophiles Ex. Thermococcus, Sulfolobus, Thermosphaera
b. Methanogens Ex. Methanobacterium, Methanococcus, Methanosarcina
c. Halobacteria Ex. Halobacterium, Halococcus, Natronomonas
d. Thermoplasma Ex. Thermoplasma

2. Cyanobacteria

They are the filamentous, oxygenic photosynthetic bacteria. They have special cells called heterocyst in which nitrogenase enzyme is present. The nitrogenase enzyme is responsible for fixing atmospheric N2 into ammonia. Some cyanobacteria are single celled and some are non-heterocyctous. Important groups are as follows:

1. Single celled cyanobacteria Ex. Chrococcus, Gleotheca, Gleocapsa
2. Filamentous non-heterocystous cyanobacteria Ex. Oscillatoria, Lyngbya
3. Filamentous heterocystous cyanobacteria Ex. Anabaena, Nostoc, Tolypothrix

3. Anoxygenic phototrophs

Single celled, sulphur required bacteria. They use H2S as electron donor. Ex. Green sulphur bacterium Chlorobium.

Vol 2. Proteobacteria

This volume has gram negative bacteria. They are further divided into 5 subgroups as α, β, γ, δ and ε. They contain medically, industrially and agriculturally important bacteria.

S.No.	Important Bacteria	Characters	Example
α Proteobacteria			
1	Purple bacteria	Anoxygenic Photosynthetic-sulphur bacteria	*Rhodospirillum, Rhodobacter, Chromatium*
2	Associative Nitrogen fixing bacteria	These bacteria present in the rhizospher of graminaceous plants and symbiotically fix atmospheric nitrogen	*Azospirillum*

3	Symbiotic Nitrogen fixing bacteria	Form nodules in legume roots and fix atmospheric nitrogen. Some form galls in the roots	*Rhizobium,* *Bradyrhizobium* *Agrobacterium*
4	Free living Nitrogen fixing bacteria	Present in the soil as heterotrophs- use verity of carbon sources in soil and fix atmospheric nitrogen	*Azotobacter,* *Beijerinkia*
5	Pseudomonas group	Some are plant growth promoting Rhizobacteria some are pathogens some produce alcohol	*Pseudomonas* *Xanthomonas* *Zymomonas*
6	Rickettsia	Endoparasites	*Rickettsia*
7	Sulphur oxidizing bacteria	Uses S as electron donor – Chemolithotrophs – strict aerobes	*Thiobacillus*
8	Acetic acid producing bacteria	Fermentative bacteria	*Acetobacter,* *Gluconobacter*
9	Budding bacteria	Reproduction by budding like yeast	*Caulobacter*
10	Hydrogen bacteria	Hydrogen producing bacteria	*Alkaligenes*
β Proteobacteria			
1	Nitrifying bacteria	Chemolithotroph – strict aerobe- soil bacteria – important form N cycle	*Ammonia to nitrite – Nitrosomonas* *Nitrite to nitrate – Nitrobacter*
2	Neisseria & relatives		Neisseria
3	Spirillum	Aerobes & facultative aerobes	*Spirillum sp.*
4	Sheathed bacteria		*Sphaerotilus*
γ Proteobacteria			
1	Purple sulphur bacteria	Anoxygenic photosynthetic- sulphur bacteria	*Thiobacillus,* *Thiospirillum*
2	Methylotrophs	Uses methane and methanol as carbon source	Methylomonas, Methylobacter, methylococcus
3	Coliforms	Present in the intestinal track of mammals	*Escherichia,* *Salmonelia*
δ Proteobacteria			
1	Sulphur reducing bacteria	Anaerobes – use S as terminal electron acceptor	*Desulfovibrion,* *Desulfomonas*

2	Glding bacteria	Gliding movement	*Myxobacteria*
3	Vibrio group	Most are pathogenic	*Vibrio, Erwinia*
ε Proteobacteria			
1	Camphylobacter,		
2	Thiovulum,		
3	Thiomicrospora,		
4	Heliobacter are		
5	Common examples		

Vol 3. Low G + C gram positives

S. No.	Group	Characters	Example
1	Clostridia group	Strict anaerobes – mostly fermentative nutrition- few thermotolerant – endospore producers	*Clostridium, Thermoanaer Obacterium,* Thermoanaerobium
2	Mycoplasma group	Absence of cell wall	*Mycoplasma, Mesoplasma, Spiroplasma*
3	Bacilli and Lactobacilli group	Lactic acid producing bacteria – endospore producers – aerobes – aerotolerant – fermentative nutrition	*Leuconostoc, Lactococcus, Streptococcus*

Vol 4. High G+C gram positives

S. No.	Group	Characters	Example
1	Actinomycetes	Filamentous – sporangiospores – conidiospores – soil habitat – antibiotics producers Symbiotic with cauarina – form root nodules – N2 fixation	*Actinomyces, Nocardia, Sreptomyces* *Frankia*
2	Mycobacterium group	Presence of mycolic acid in the cell wall – acid fast staining – human pathogens	*Mycobacterium lepri*
3	Corynebacterium group	Human pathogens	*Corynebacterium diptheriaea*

Vol 5. Plancomyceters, Spirocheter, Bacteroides and Fusobacteria

S. No.	Group	Characters	Example
1	Chlamydia group	Obligate parasites to man, animal and birds	*Chlamydia*
2	Bacteroides	Obligate anaerobes	*Bacteroides*
3	Spirochete	Gram negative – flexile- endoflagella presence	*Spirochete, Leptospira*

Chapter - 6

Fungi

Eukaryotic, spore-producing, achlorophyllous organisms with absorptive nutrition that generally reproduce both sexually and asexually and whose thallus is usually filamentous, branched somatic structures known as hyphae, typically surrounded with a cell wall.

Mycology – Study of fungi

The fungi can be divided into two groups namely

True fungi: Chytridiomycota, Zygomycota, Ascomycota, Basidiomycota

Additional phyla: Myxomycota, Dictyosteliomycota, Acrasiomycota, Plasmodiophoromycota, Oomycota, Labyrinthulomycota, Hypochitridomycoya.

Economic Importance of Fugi

Following are the benefits of the fungi:

S. No.	Benefits	Process	Example
1	Biotran-formations	Conversion of carbohydrate rich compounds to value added products like alcohol, steroids, brad etc.	Yeast (alcohol & bread)
2	Useful metabolites	Antibiotics, plant growth regulators, hormones etc.	Penicillium (Penicillin) Fusarium (Gibberellic acid)
3	Enzymes	Industrial production of enzymes	Aspergillus (Amylase)

4	Biomass	Food and feed additives	Agaricus (Food) Yeast (Feed additive)
5	Biological control	Control the insects, other fungal pathogens, nematodes etc.	Tichoderma (Against soil plant pathogenic fungi)
6	Soil fertility	Nutrient cycle, decomposition of organic matter	Most of the soil fungi

Following are the harmful effects of fungi:

S. No.	Harmful effect	Cause/Symptom	Example
1	Food spoilage	Make the foods unable to use	Aspergillus (Stored grains) Mucor (Bread)
2	Mycotoxins (referred as the toxins produced by fungi on certain plant materials that will either consume directly as food or feed)	Causes diseases, tumors to the host animals	Aspergillus flavus (Groundnut)
3	Human & animal diseases	Ill health, diarrhea, skin diseases	
4	Plant diseases	Root rot, rust, powdery mildew, blight, leaf spot, smut, etc.	Phytophthora infestans (Late blight of potato)

General characters of fungi

» Heterotrophic organisms devoid of chlorophyll

» The resemble plant by

- Cells wall
- Non-motile
- Reproduce by spores

» Separate from plants

- No leaf, stem, root
- No chlorophyll

- Carbohydrate storage is glycogen instead of starch
- Filamentous

» Hypha: Body of fungi; microscopic tubular thread like hyphae branch in all directions. Collectively mass of hyphae is referred as mycelia (Singular : mycelium). Some don't have these structures. Ex. Yeast

» They use variety of carbon sources like cellulose, starch, hemicelluloses etc.

» Temperature optima: 25-30^{0}C C; Max: 40^{0}C Min: 10^{0}C some thermophilic fungi prefer to grow at 50^{0}C.

» pH optima: 4-7

» Oxygen: Majority of them aerobes; Some facultative anaerobes (Ex. Yeast), and some obligate fermentative type (Ex. Chytridomycota)

» Light: Not required

Dimorphic: Some organisms can exist as either hyphae or single-celled. Ex. Disease-causing fungi of humans and animals. They exist as hyphae when outside the host and in the host, as single-celled. They are referred to as dimorphic fungi.

Morphology of fungi

Hyphae: thin, transparent, tubular wall – filled with protoplasma.

The cross walls are referred as septa (Septum).

If septum is present, the hyphae are referred as septate hyphae and absence of septa, aseptate/nonseptate/coenocytic hyphae.

Cell wall

The cell wall of fungi is made up of cellulose, glucans, chitins etc.

Glucan: glucose with ʊ - 1, 3; ʊ - 1, 6; β – 1, 6 and β – 1, 4 linkages

Chitin: b – 1, 4 linked N- acetyl glucosamine

Oomycetes- cellulose is the major compounds of cell wall

Normally, the hyphae/mycelium are unnoticed in underground nature, which cannot be seen by the eyes. The mycelium is always embedded in the soil/host plant or animal. But sometimes they become organized to form large structures to see with the naked eye, which are referred to as special vegetative structures. Following are some of the modifications of vegetative structures of saprophytic fungi.

1. **Stroma:** compact, somatic structure much like a miniature mattress or cushion formed are referred as stroma.
2. **Sclerotia:** Hard resting bodies resistant to unfavorable conditions E: Sclerotium sp. and Claviceps purpurea
3. **Rhizomorphs:** Thick strands of hyphae, consolidated tissue like are referre as rhizomorphs. They become dormant at adverse condition and useful fc nutrient absorption.

Following are the modification of vegetative structures of parasitic fungi.

1. **Haustoria:** The fungus penetrates the host cells by branched hypha an produce bulbous structures are called as haustoria. This structure is used t absorb the nutrients from host plant cell. Ex. Rust fungi.
2. **Appressoria:** Infection structures formed at the tip of germ tubes or hypha outside the host. The appressoria is the bearing structure of parasitic fung Ex. Puccinia, Peranospora

Nutrition

Fungi perform the absorptive type of nutrition. They derive food by four differer ways as follow:

1. **Saprophyte:** Uses dead organic matter to derive food – Heterotrophs Ex. Muca Aspergillus, Penicillium.
2. **Parasites:** Uses living tissues to derive food. The parasitic fungi were furthe divided into three groups as follows based on their mode of nutrient absorptio from host.
   - Perthotroph/Necrotroph: Use enzyme or toxin to kill the host cell and deriv the food from dead cells.
   - Biotrophs: Produce special vegetative structures (haustoria) and absorb th nutrients from live cells. They are ecologically obligate parasites.
   - Hemibiotrophs: Initially require living host and later uses the dead cells E: Colletotrichum lindemuthianum.
3. **Symbiosis:** Derive food from the host by beneficial each other. Ex. Lichens Algae and fungi sybmiosis; Mycorrhiza – Higher plant roots and fungi symbiosi
4. **Predation**: Animal trapping fungi capture the small animals like worms, protozo etc. and kill the cells and derive their food. Ex. Arthobotrys, Dactylella

Ecological groups

The fungi can be divided into many groups based on the ecological conditions

S. No.	Group	Habitat	Example
1	Soil fungi	Soil	Penicillum, Aspergillus
2	Lignicolous fungi	Woody tree barks where rich of lignin	Polystictus
3	Entomogenous fungi	Insects	Endomopthora
4	Coprophilous fungi	Cowdung	Pilobolus
5	Aquatic fungi	Water	
6	Cellulolytic fungi	Cellulose rich soil	Chaetomium
7	Dermatophytes	Skin	Trichophyton

Reproduction of fungi

Fungi can reproduce by vegetative, asexual or sexual reproduction.

Vegetative reproduction

1. Fragmentation: The hypha fragment to many cells and each cell will raise as new individual. Ex. Yeast.
2. Fission: As bacteria, binary fission gives two new daughter cells. Ex. Yeast.
3. Budding: In yeast, the cell forms a bud like appendage, in which the divided nucleus will move and new cell formed. The detached bud will raise to new individual.
4. Sclerotia: resting vegetative structure can give the new individual.
5. Rhizomorph: also gives the new individual.

Asexual reproduction

1. Arthospores/Oidiospores: The broken hypha behaves as spores are referred as arthospores or oidiospores.
2. Chlamydospores: The cells become resting body and separate from hyphae referred as chlamydospores. When favourable condition araises, the chlamydospore will form new mycelium Ex. Fusarium.
3. Asexual spores: Sporangiospores – spores produce from sporangium (sac like organ); Conidia – Spore produced at the tip of the hyphae.

Exogenous & Endogenous spore

Spore produced inside the thallus is referred as endogenous spore (Ex. Sporangiospore) and spore produced outside the thallus is referred as exogenous spore (Ex. Conidia).

Motile & Non-motile spores

Spores with flagella are referred as Planospores. The spores will have whiplash and tinsel type flagella either or both. Ex. Zoospores.

Spores without flagella are referred as Aplanospores. Ex. Sporangiospores, conidia.

A part from these basic type of spores, pynidia, pycnia, acervuli, sporodochia are some other asexual spores produced by Basidiomycetes fungi.

Sexual reproduction

Sexual reproduction of fungi involves the following sequence of processes.

1. Plasmogamy: Union of protoplasm of two thallus or gametes.
2. Karyogamy: Union of two nucleus
3. Meiosis: Cell division to maintain the n level of chromosome.

Generally, all the thallus are at 'n' level of chromosome, referred as haplophase and before meiosis, the chromosome level in '2n' which is referred as diplophase.

After meiosis spores were formed, which vary with different classes of fungi. The sexual spores are Oospores; Zygospores; Ascospores and Basidiospores.

The sex organs of fungi are referred as Gametangia / gametangium

The cell / nucleus involved in the sexual reproduction is gamete and the fusion of two gametes lead to sexual reproduction i.e., plasmogamy, karyogamy and meiosis.

Isogametangium & Isogametes

Male & female sex organs and sex cells cannot be differentiating morphologically.

Anisogamethangium & Anisogametes

Male and female sex organs and sex cells (gametes) structurally differ each other is referred as Anisogametangium and anisogametes.

Male gametangium – Anthridium

Male gamete – Anthrozoid

Female gametangium – Oogonium, Ascogonium

Female gamete – Oosphere

Union of two isogametes is referred as isogamy and union of anisogameters is referred as anisogamy.

1. Monoecious / Hermaphrodite: Male and female sex organs present is same thallus
2. Dioecious: Male and female are present on different thalli.

Homothallic fungi: Male and female gametangia are present in same thallus and can reproduce by itself referred as homothallic fungi. Also referred as "Self – fertile".

Heterothallic fungi: Male and female gametangia are present in different thallus and cannot reproduce by itself referred as heterothallic fungi. Also referred as "Self – sterile" – need another type of thallus for sexual reproduction.

Pseudohomothallic fungi: They are heterothallic fungi (self sterile) but their spores are self fertile referred as pseudohomothallic fungi.

Holocarpic & Eucarpic Fungi

If whole thallus / hyphae used or converted as the reproductive phase, the fungi is referred as holocarpic fungi. Ex. Yeast. Reproductive organ arise from the thallus referred as eucarpic fungi. Ex. Aspergillus. The sexual stages of these fungi are referred as teleomorph; asexual stage is anamorph and collectively referred as holomorph.

Common Sexual Reproductions

1. Planogametic copulation

Fusion of naked motile gametes is referred as planogametic copulation. Ex. Plasmodiophora "Planogametes" – motile gametes.

2. Gametangial contact

Two gametangia come in contact but not fuse. The male nucleus (gamete) migrates through a pore or fertilization tube to female gametangium and sexual reproduction takes place. Ex. Aspergillus, Peniicillium.

3. Gametangial copulation

Two gametangia (male and female) fuse and sexual reproduction takes place, referred as gametangial copulation. Ex. Zygospore of Zygomycetes – Mucar, Rhizophus.

4. Spermatization

A uninucleate nonmotile spore-like male structures acted as "male gamete" (Spermatium) (gametangium: spermagonium / antheridium) which pass through female gemetangium (Oogonium) leads fertilization. This process of sexual reproduction is referred as spermatization. Ex: Neurospora.

5. Somatogamy

The sex organs are not produced. The somatic cells (thallus) take part in the sexual fusion is referred as somatogamy. Ex. Morchella.

Parasexuality: The sexual reproduction (plasmogamy – karyogamy – meiosis) occurs in sequence but not at a specific points in the life cycle of an individual is referred as parasexuality. Ex. Aspergillus nidulans.

Fungal classification

The recent advance in fungal classification includes:

- » Recognition of artificial nature of three / five kingdom classification
- » Theory of data analysis techniques of phylogenetic systems.
- » Molecular techniques in mycology
- » Discovery of new fossils and taxa

Present classification is "Phylogenetic classification" based on evolutionary relationship The fungi can be grouped in to two groups based on their ancestors.

Monophyletic groups: Group of individuals that contain a common ancestor and all its descendants.

Polyphyletic groups: Group of individuals that donot share a common ancestor.

The fungi are divided into 3 major groups

1. Kingdom: Fungi (one monophyletic group)

Note: The common name fungi and kingdom Fungi are differentiated by writing as scientific name – italic or underline

- Phylum : Chytridiomycota
- Phylum : Zygomycota
- Phylum : Ascomycota
- Phylum : Basidiomycota

2. Kingdom – Stramenopila (one monophyletic group)

- Oomycetes
- Hyphochytriomycota
- Labyrinthulomycota

3. Protists (All are polyphyletic groups)

a. Plasmodiophoromycota

b. Dictyosteliomycota

c. Acarysomycota

d. Myxomycota

The protists are formerly referred as "Slime molds".

Kingdom: Fungi

Phylum: Chytridiomycota

- This phylum has single class – chytriomycetes.
- Only members of fungi produce motile cells at some stage of their life history.
- Asexual spores – Zoospores.
- Sexual reproduction – Gametangial copulation
- Coenocytic structure of hyphae with well developed mycelium.
- Eucarpic.
- Rhizoids / rhizomycelium – well developed used for nutrition absorption and for anchor
- These fungi are referred as "Chitrids"
- Occur in aquatic, soil, cattle rumans etc.
- Many orders, families are present.

Type genera: Chytriomyces, Allomyces, Synchytrium.

Phylum: Zygomycota

This phylum includes two classes namely Zygomycetes and Trichomycetes.

Class: Zygomycetes

- Characterized by thick walled resting spores called "Zygospores" – develope through zygosporangium.
- All individuals produce this sexual spore.
- Morphological variation is there.
- Coenocytic mycelium.
- Asexual reproduction – sporangiospores (Aplanospores)
- Sexual reproduction – Zygospores
- Gametangial copulation leads to formation of Zygospore, which is resting spore. When favorable condition, zygospore germinate gives th sporangiospores.
- Disease causing organisms
- Food borne pathogens
- Food spoilage
- Type genera: Mucar, Rhizopus, Entomophthora.
- Important orders are Mucarales; Entomophthorales; Zoopagales.

Class: Trichomycetes

- Obligate associated with insects, millipedes, arthropods.
- Symbiotic association occurs.
- Most of them grown internally within the gut of these animals referred a 'holdfast'
- Asexual reproduction
- Amoeboid cells.
- Arthrospores
- Sporangiospores.
- Trichospores: These are exospores with single uninucleated spores exclusivel present in this class fungus.
- Sexual reproduction: Zygospores.

Type genera: Harpella, Smittium, Amoebidium, Asellaria, Enterobryus.

Phylum: Ascomycota

The characters of Ascomycota fungi are

- » Septate mycelium
- » Sometimes dikaryotic mycelium (2 nuclei per cell)
- » Ascospores – sexual spores – 4 or 8 ascospores in a sac – called "Ascus"
- » Isogametangial contact, anisogametangial contact, spermatization and somatogamy are the common sexual reproduction methods.
- » Fruiting body of Ascospores is "Ascocarp"
- » Few don't produce ascocarp Ex. Yeast.
- » Asexual spores: Conidia / sporangiospore – Aplanospores
- » Varieties of shapes of conidia are present.
- » Few individuals have only asexual stage and few have both sexual and asexual stage.

The fungi which have both sexual and asexual stage are referred as perfect fungi and the fungi having the asexual stage alone are referred as imperfect fungi.

The structure of ascocarp varies with three different types:

1. Cleistothecium: Closed ascocarps are called as cleistothecium.
2. Perithecium: Closed ascocarp but a pore or opening is present to facilitate the release of ascospores are called as perithecium.
3. Apothecium: Opened ascocarp is referred as apothecium.

Different Classes of Ascomycota

1. Asexual ascomycetes: Deuteromycetes / Imperfect fungi only asexual reproduction occurs by means of sporangia / conidia. Ex. Penicillium, Paecilomyces, Metarrhizium.
2. Archiascomycetes: Saprophytic & parasitic fungi. Ex. Taphrina, Schizosaccharomyces.
3. Order: Saccharomycetales. Yeast: 4 Ascospores per ascus; Ex. Saccharomyces, Candida, Kluyveromyces.

4. Order: Eurotiales (filamentous ascomycetes). Ascospores with cleistothecium type ascocarp Ex. Aspergillus, Paecilomyces, Talaromyces.
5. Ascomycetes with perithecia (Pyrenomycetes): Perithecium type ascocarp. Ex. Neurosposa, Fusarium, Claviceps
6. Ascomycetes with Apothecium (Discomycetes). Ex. Lichens, Mycorrhiza

Lichens: Association of fungus with algae / cyanobacteria in which a single thallus formed are referred as lichens. The fungal component is referred as mycobiont and algal component is photobiont. Ex: collema, Parmelia.

Ecological groups of lichens:

Rock surface	:	Saxicolous lichens
Inside rock	:	Endolithic lichens
Soil	:	Terricolous lichens
Bark	:	Corticolous lichens
Leaves	:	Folicolous lichen

Mycorrhiza (Myco – fungi, rhiza – roots)

The association between fungi and higher plant roots is referred as mycorhhiza. They are divided into two groups as endomycorrhiza, which penetrate the root cell and symbiotically beneficial with roots; ectomycorrhiza, which colonize on the surface of the higher plant roots and symbiotically beneficial with roots. Ex. Glomus, Gigapora – Endomycorrhiza; Amanita, Bolytus – Ectomycorrhiza.

1. Ascomycetes with Ascostromata: Production of asci in the stroma (cushion like structure) referred as ascostroma is the typical character of this class fungi. Ex: Venturia, Podonectria.
2. Other filamentous ascomycetes – Powdery mildew disease causing fungi. Ex. Erysiphae, Leveillula.

Phylum: Basidiomycota

The characters of basidiomycete's fungi are as follows:

» Sexual spore: Basidiospores

» Spore forming structures: Basidium

» Fruiting body: Basidiocarp

- 4 basidiospores per basidium formed
- There are three different stages of life cycle such as a. primary mycelium – septate – single nucleated b. secondary mycelium – septate – heterokaryons stage and c. tertiary mycelium in which basidiocarp formation
- Asexual reproduction: Oidium / oidiospores

Following are the sub groups of basidiomycetes

1. Order: Agaricales (Mushrooms) Ex. Agaricus, Coprinus.
2. Class: Gasteromycetes: Puffballs, earth stars, Birds nest fungi
3. Order: Aphyllophorales: Polyporus, Chantarelles, tooth fungi, coral fungi, corticoids.
4. Order: Uridinales: rust fungi, Puccinia, Elromyces.
5. Order: Ustilaginales: Smut fungi Ustilago, Tilletia, Urocystis.
6. Other basidiomycetes Ex: Tremella.

Kingdom: Stramenopila

Phylum: Oomycota

- Class: Oomycetes
- Sexual: Oospores
- Asexual: Zoospores
- Type genera: Pythium, Phytophthora, Perenospora, Plasmopara, Albugo

Phylum: Hypochitridomycota

- Holocarpic fungus
- Type genera: Rhizidiomyces apophystaus

Phylum: Labyrinthulomycota

- Net slime molds

Protists

These groups of organisms are not true fungi. They don't have mycelia structure as that of true fungi. They are phylogentically primitive to fungi and have the characters of fungi and protozoa.

Phylum: Plasmodiophoromycota

- Endoparasitic slime molds
- The thallus is referred as plasmodia. They are acellular, mass of protoplasm in which the nuclei are present. Ex. Plasmodiophora.

Phylum: Dictyosteliomycota

- Cellular slime molds Ex. Dictyostelium

Phylum: Myxomycota

- True slime molds: Myxcomycetes skemonitis.

Chapter - 7

Algae

lgae are a large and diverse group of eukaryotic organisms that contain chlorophyll nd carry out the oxygenic photosynthesis.

Study of algae is referred as phycology or algalogy

They vary from cyanobacteria or blue green algae by their eukaryotic nature.

)ccurrence

he algae are mainly present in aquatic ecosystems. They are present in

» Oceans, sea, salt lakes, fresh water lakes, ponds and streams.
» Damp soils, rocks, stones, tree barks etc.
» Small aquatic free floating microscopic algae in water are referred as planktons or phytoplanktons.
» Some are present as parasites of plants.

conomic Importance

ollowing are the beneficial effects of algae:

» The algae are the primary producer of food chain of aquatic ecosystem.
» They are also the CO_2 fixers.
» Algal cell wall contains several economic compounds such as alginic acid, agar and carrageenan.

Agar: is a polysaccharide obtained from the red alga – Gelidium. It is used as solidifying gent for media preparation in microbiology labs; cheese, jelly preparations and as packing for canned foods.

Alginic acid: is obtained from brown algae Macrocystis, Agarum, and Laminari It is useful in the smooth consistency of ice cream, tooth impressions of dental lab jelly preparations, and immobilization matrices using calcium alginates.

Carrageenam: Obtained from Chondrus, Gigartina. It acts as a stabilizer of food and ice cream, a binder of toothpaste, a finishing compound of the textile and pap industry, a thickening agent of shaving creams, soaps, etc.

- » Diatoms are the group of algae which contain silica in cell wall. Th diatomaceous earth is used for filter preparations.
- » Porphyra (a red algae) used as food (Japanese uses this alga for No preparations)
- » Chlorella (green algae) used as feed additive for domestic animals.
- » Spirulina (green algae) has many medicinal properties
- » The different group of these algae is used for pigments production such a carotenoids, xanthophylls, phycocyanins etc.

Following are some harmful effects of algae:

- » Prototheca alga produces bursitis to mammals and human (Inflammatio of joints)
- » Cephaleurous parasiticus causes red rust disease to tree crops such as te coffee, mango and sapota.
- » Aquatic algae produce neurotoxins which are of highly poisonous (Ex Gymnodium).

Morphology

Wide range of size and shapes are occurred in algae.

1. Unicellular: single cells, motile or nonmotile Ex. Chalamydomonas, Chlorell
2. Colonies: Assemblage of individual cells with variable or constant number o cells that remains constant throughout the colony life Ex. Diatoms, Volvox
3. Filaments: daughter cells remain attached after cell division and form a cel chain; adjacent cells share cell wall (distinguish them from linear colonies Ex. Spirogyra, Spirulina.

4. Coenocytic forms: one large, multinucleate cell without crosswalls Ex. Vaucheria
5. Parenchymatous and pseudoparenchymatous algae: mostly macroscopic algae with tissue of undifferentiated cells and growth originating from a meristem with cell division in three dimensions; pseudoparenchymatous superficially resemble parenchyma but are composed of appressed filaments Ex. Porphyra

» Eukaryotic nature

» Presence of thin and rigid cell wall

» Presence of chloroplasts

Cell wall

» The algal cell wall is primarily made up of cellulose fibres. But, the cellulose fibres are modified by additional polysaccharides like xylans, mannoses, alginic acids etc.

» Some algal cell walls deposited with calcium carbonate, referred as calcarious or coralline algae.

» Some algae, the cell wall is absent, Ex. Euglena

» Silica deposited cell wall Ex. Diatoms. They are extreme resistance to decomposition.

Nutrition of Algae

» Photosynthesis is the primary mode of nutrition (C uptake as CO2)

» Photosynthetic algae are mostly considered as photoautotrophic.

» Numerous phototrophic algae can, however, take up dissolved organic matter or engulf bacteria and other algal cells as particulate prey; they are referred to as mixotrophic.

» Some algae cannot synthesize essential vitamins (biotin, thiamine, cobalamin = B12) and have to import them: referred as auxotrophic species.

Motility

Presence of flagella for motility for few motile algae Ex. Euglena, Chlamydomonas

Dianoflagellates have two flagella with different lengths.

Pigments

Since algae are the photosynthetic organisms, their light harvesting system, by pigments varies among different group of algae. Basically the pigments are grouped into following categories:

1. Chlorophyll – a, b, c, d, and e – lipid soluble
2. Carotenoids – Carotenes and xanthophylls – lipid soluble
3. Biloproteins (Phycobilins) – Phycocyanin and phycoerythrin – water soluble

The concentration of these pigments varies with different algae leads to appearance of particular colour. Ex. Chlorophyll rich leads to green colour; phycoerythrin rich leads to red colour; and so on.

Reproduction

1. **Asexual Reproduction:** One individual can produce copies of itself.
   - » Binary fission: many unicellular algae, longitudinal or transverse cell division.
   - » Fragmentation: colonies or filaments break into two to several pieces that continue to grow.
   - » Akinetes: Enlarged vegetative cell with thick wall and storage products; survival of harsh conditions rather than additional copies.
   - » Zoospores / Aplanospores: Zoospores are flagellated reproductive cells from which a new individual/colony can grow; sometimes the spores begin to develop within the mother cell and lack flagella: aplanospores; aplanospores can develop into zoospores.
2. **Sexual Reproduction:** Plasmogamy, the fusion of haploid cells (gametes), precedes karyogamy (nuclear fusion) to form a diploid zygote. These are the principle events of sexual reproduction.
   - » Gemetes look like vegetative cells or very different.
   - » Isogamy: both gemetes look identical.
   - » Anisogamy: male and female gametes differ morphologically
   - » Oogamy: One gamete is motile (male), one is nonmotile (female)
   - » Monecious: both gametes produced by the same individual.
   - » Diecious: male and female gametes are produced by different individuals.
   - » Homothallic: gametes from one individual can fuse (self-fertile).

» Heterothallic: gametes from one individual cannot fuse (self-sterile).

» Gametophyte: typically haploid, produces gametes by mitosis.

» Sporophyte: typically diploid, produces spores by meiosis.

All types of reproductions studied in fungi will be applicable to algae too. Note: refer fungi for further details

Taxonomic Groups

Phylogeny of Algae

» Algae did not evolve from one single, common ancestor but evolved at different times from different ancestors. (Polyphyletic group) so, Algae are therefore, not monophyletic

» Some algae are closer related to heterotrophic flagellates or ciliates than to other algae.

» Eukaryotic algae: posses a double membrane enclosed nucleus and usual other membrane-enclosed organelles (mitochondria, chloroplasts)

» Origin of organelles: symbiotic enclosure of free living bacteria or cyanobacteria, horizontal transfer of genes from endosymbionts to host nucleus.

The pigments are the main taxonomic character for grouping the algae in to different phyla. The following table shows different groups of algae with suitable examples.

Taxonomic group	**Pigments**			**Examples**
	Chlorophyll	**Carotonoids**	**biloproteins**	
Rhodophycophyta (Red algae)	Chl a,d	β-carotene Zeaxanthene	Phycoerythrin Phycocyanin	*Polysiphonica* *Gelidium*
Xanthophycophyta (Yellow green algae)	Chl a,d	β-carotene heteroxanthene	-	*Vaucheria*
Chrysophycophyta (Golden algae)	Chl a,d	β-carotene Fucoxanthene	-	*Ochromonas*
Phaeophycophyta (Brown algae)	Chl a,d	β-carotene α-carotene Fucoxanthene Violaxanthin	-	*Kleps, Sorgassum natans*
Bacillariophycophyta (Diatoms)	Chl a,d	β-carotene	-	*Diatoms*

Chlorophycophyta (Green algae)	Chl a,d	β, γ-carotene	-	*Chlorella, Chlamydomonas, Spirogyra*
Cyptophycophyta (Cryptomonads)	Chl a,d	α-carotenes	Phycoerythrin Phycocyanin	*Cryptomonas*
Pyrrophycophyta (Dianoflagellates)	Chl a,d	β-carotene, Peridinin	-	*Noctiluca*
Euglenophycophyta (Euglenoids)	Chl a,d	-	-	*Euglena*

Lichens

Association of fungi with algae or cycnobacteria in which a single thallus formed are referred as lichens.

Fungal component is referred as mycobiont and algal component is photobiont

Thallus

Top layer of tightly woven mycelium, below which the photosynthetic cells and followed by another layer of fungus formed a complex structure which can not be distinguished the algae and fungi.

Fungi: Ascomycetes and few basidiomyceters; Algae – green algae.

About 18000 species of lichens have been identified. Based on their structure they were divided into three groups as:

1. Crustose: flat apprised
2. Foliose: leaf like
3. Fructicose: shrub like

Reproduction: by means of sordia (knot of fungal hyphae with few algal spores are referred as sordia)

Economic importance of lichens

1. Production of unusual compounds such as phenolics etc.
2. Commercial production of litmus pigment indicator
3. Indicator system form pollutants.

Chapter - 8

Protozoa

Proto – first; Zoon – animal

Protozoa are single celled eukaryotic organisms, which lack cell and ability to move at some stage of their life cycle.

They are referred as the first animals.

There are about 65000 described species grouped into 7 phyla. Among them, 50 per cent are not available as living forms (as fossils) and out of remaining 50%, [illegible]2,000 species are free-living and 10,000 species are parasitic forms.

The protozoa are distinguished from other organisms by following characters

- From bacteria- Eukaryotic nature and greater size
- From algae – Lack of chlorophyll and cell wall
- From fungi – Motility and lack of cell wall
- From slime molds (Protists) – Lack of fruiting body

Occurrence of Protozoa

- Present in all moist habitat
- Sea, soil and fresh water are common habitat
- Also present as parasites in higher animals

Ecological Groups

Protozoa may be divided into three groups based on their ecological conditions namely,

1. Free-living: Occur in salt water, fresh water, sand, soil and decaying organ matter
2. Symbiotic forms: Occur in guts of termite and help in the digestion of wooc materials. Ex. Trychonympha protozoa present in the guts of termites hel to digest the cellulose of woods.
3. Parasitic forms: Occur in the body of higher animals and human and caus diseases.

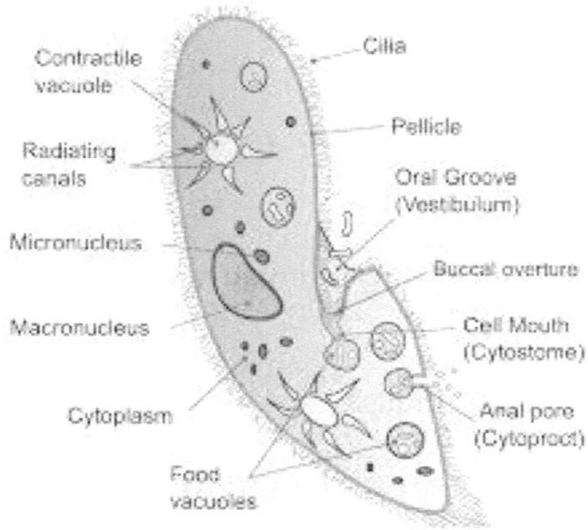

Fig. 8.1: A Few Well- known Protozoa (Source: https://byjus.com/neet/protozoa

Morphology

Protozoa exhibit a wide variety of morphologies. There is no definite shape c morphology that would include a majority of protozoa. Shapes range from amorphou and ever-changing forms of amoeba to relatively rigid forms. Size and shape of thes organisms show diversity.

Shape varies from amoeboid to all possible shapes.

Size ranges from 200 to 600 mm diameter protozoa are available.

Plasmodium – 2-5 mm; Paramecium – 200 – 500 mm; Spirostomum – 3 mm

Like eukaryotic cells, the protozoan cells consist of cytoplasm, nucleu endoplasmic reticulum, and all regular cell organelles. Several protozoa expres photosynthetic pigments thus colored. (Ex. Euglena) Many protozoa exhibit comple life cycles with multiple stages (e.g., Plasmodium).

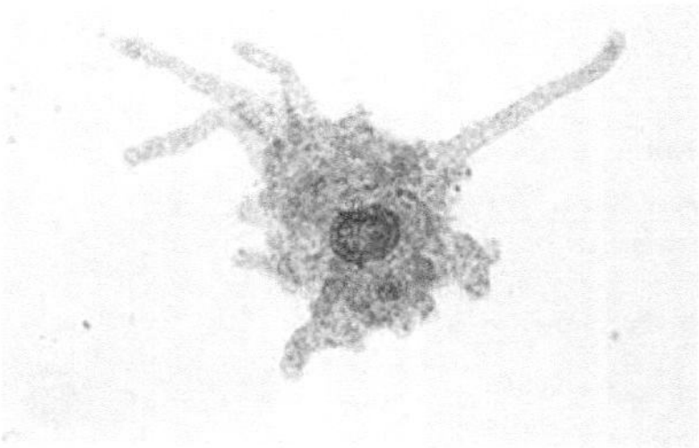

Fig. 8.2: Amoeba (Source: livesciences.com)

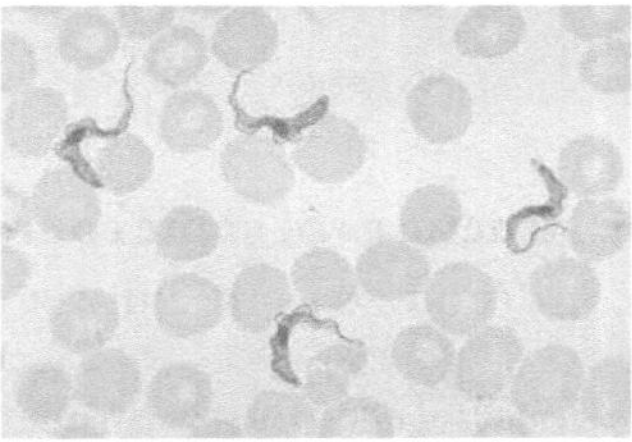

Fig. 8.3: Trypanosome (Source: Wikipedia.org)

Cyst: Many protozoa form resistant cysts at certain times of their life cycle. They are able to survive in adverse environmental conditions such as desiccation, low nutrient supply, and even anaerobiosis. In parasitic protozoa, the cyst stage is often used to transmit from one host to another.

Locomotive Organs

The motility is the key characteristic of protozoa, and the mode of motility varies with organisms. The following are the possible ways by which the protozoa motility occurs.

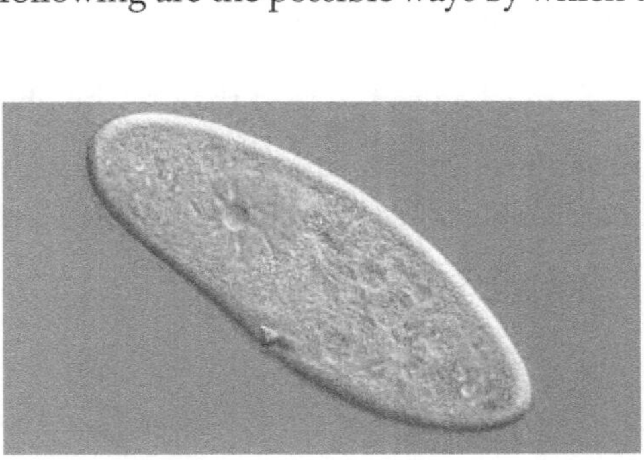

Fig. 8.4: paramecium (Source: livesciences.com)

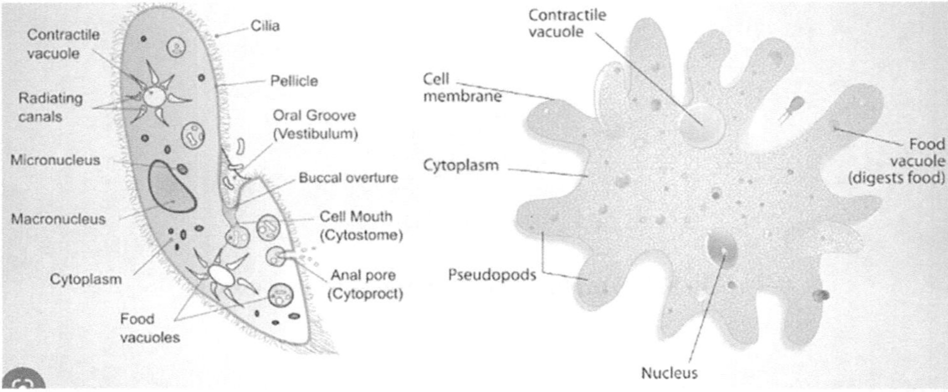

Fig. 8.5: Fine structures of some common protozoa
(https://microbiologynote.com/protozoa-structure-and-locomotor-organelles/)

1. Pseudopodia (False foot): Temporary projection of part of cytoplasm of those protozoa which don't have rigid cells are referred as pseudopodia. Ex. Amoeba. The pseudopodia also helpful to capture food through engulfing.
2. Flagella: Fine filamentous extension of cell similar to that of bacteria. The number varies from 1 to 8. Ex. Euglena.
3. Cilia: Fine short thread like extension from cell used for locomotion and ingestion of food. Ex. Paramecium
4. Gliding motility: Ex. Sporozoa.

Feeding

The ingestion is the primary mode of feeding habit of protozoa.

1. Pseudopodia of amoeba used to feed the prey by phagocytosis
2. Cytosome or oral groove: An opening of ciliate protozoa through which food is ingested. It ranges from simple round opening to slit like structures.

   Feeding by Pseudopodla Ex. Amoeba

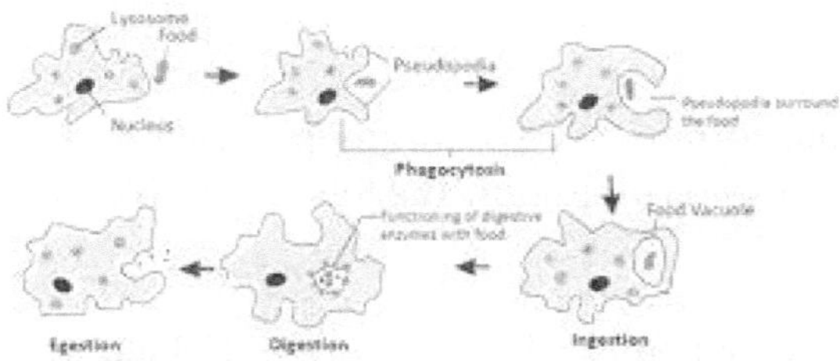

Fig. 8.6: Feeding by Pseudopodla (Source https://www.embibe.com/exams/feeding-digestion-in-amoeba/)

Feeding by oral groove (Cytosome) Ex. Paramecium

Protective structures

Since the protozoa are habitat of aquatic nature, they are provided with protective structures. They are:

1. Mucocysts: Certain ciliates secrete a mucilage substance from vesicles called mucocysts.
2. Trichocysts: Expulsion of harpoon like structures (spear like missile structures) which will attack the enemies.
3. Toxicysts: Thread like tubular structure of ciliate protozoa excrete toxins which used to paralyze and capture the prey.

Reproduction

As a general rule, protozoa multiply by asexual reproduction and few have sexual reproduction too.

Asexual reproduction

1. Binary fission: Ex. Paramecium, Amoeba
2. Multiple fission: Single cell divided into many cells Ex. Plasmodium.
3. Budding: Ex. Suctoria

Binary fission of Amoeba

Sexual reproduction

1. Sexual fusion of two gametes: Ex. Amoeba

2. Conjugation: Transfer of genetic material from one organism to anothe through cilia. Ex. Paramecium and Euplotes.

Taxonomy

The protozoa are grouped into 5 different groups based on the morphological ecological and evolutionary relationships. The following table shows the characters of individual groups.

S. No.	Group	Common name	Example	Habitat	Common disease
1	Mastigophora	Flagellate	Trypanosome leishmania	Fresh water, animal parasites	African sleeping sickness
2	Euglenoids	Phototrophic flagellates	Euglena	Fresh water and marine	-
3	Sarcodina	Amoebas	Amoeba	Animal parasites	Amoebiosis
4	Ciliophora	Ciliates	Paramecium	Animal parasites and rumens	Dysentery
5	Anicomplexa	Sporozoans	Plasmodium	Animal and insect parasites	Malaria

Economic importance of protozoa

1. Protozoa are as link of food chain of aquatic and soil ecosystem.

   Phytoplankton – Zooplankton – Primary consumers – Secondary consumer The protozoa are the zooplanktons which are referred as secondary producers
2. Ecological balance of microbial communities in aquatic and wet land ecosystem was maintained by protozoa. They eat the prokaryotes so as to maintain their population in an ecosystem.
3. Protozoa plan a major role in waste water treatment to remove the solid wastes (Ex. Anaerobic protozoa – Metopus, Sporodinium and aerobic protozoa – Bodo Paramecium).
4. Diseases to human beings Ex. Malaria – Plasmodium vivax; P. malariae and P falciparum; African sleeping sickness – Tryphanosoma leishmania; Ambebiosis – Amoeba and Entamoeba.
5. Act as molecular tool to research on animal cell behaviors.

Chapter - 9

Viruses

A virus can be defined as a genetic element containing either DNA or RNA that replicates in host cells and is characterized by having an extracellular state.

Viruses can use the metabolic machinery of the host cell and can modify the genetics of the host cell. A virus has both extracellular and intracellular states.

Extracellular State

Outside the host cell, the virus is a minute particle containing nucleic acid surrounded by protein, which is referred to as virions or virus particles. Virions are inert and have no biosynthetic and metabolic functions. The extracellular state, or virus particle, is the perfect stage of the virus to study the morphology and chemistry.

Intracellular State

The active state in which the virus replicates in the host cell. When the virus nucleic acid is introduced to the host cell and replication starts, the process is referred to as infection. The cell in which infection of virus and replication takes place is referred to as the host cell.

Chemistry of Virus

All the living cells have double-stranded DNA as their genetic material. In contrast, viruses can have either DNA or RNA as their genetic material, and it can be either single-stranded or double-stranded. The third group of viruses, which contains RNA as a genome but replicates by DNA intermediates, uses both DNA and RNA as their genetic material in their reproductive stage.

The viruses can be divided into following groups based on their nucleic acid ar their form (single/double strand).

- » DNA containing virus
  - Single strand DNA virus
  - Double strand DNA virus
- » RNA containing virus
  - Single strand RNA virus
  - Double strand RNA virus
- » RNA – DNA virus

Viruses are also divided into three major groups based on their host as 1. Pla viruses 2. Animal viruses and 3. Bacteriophages.

Structure of Virions

The structure of virus varies to spherical, rod, complex structures and the size virus is in nanometers (nm). The size ranges from 20 to 200 nm.

Ex. Small pox virus – 200 nm (smaller than smallest bacterium – Nanobacterium

Polio virus – 28 nm (about the ribosome size).

The genome (genetic material in terms of base pairs of either DNA or RNA size of each virus also varies.

- » Herpes virus – ds DNA – 150 kbp (1 kbp = 1000 bp)
- » Polio virus – ss RNA – 7 kbp.
- » Cow pea mosaic virus – ss RNA – 9 kbp.
- » Lambda (λ) Bacteriophage – ds DNA – 48 kbp.

The structure of virions is quite diversified. The nucleic acid of a virion is alway located within the particle and surrounded by a protein coat called a capsid. (Protei coat/shell/capsid are often used interchangeably).

The protein coat is always formed of a number of individual subunits referre to as capsomeres. Capsomeres are arranged in a precise and highly repetitive patter around the nucleic acid. These morphological subunits can be seen under an electro microscope.

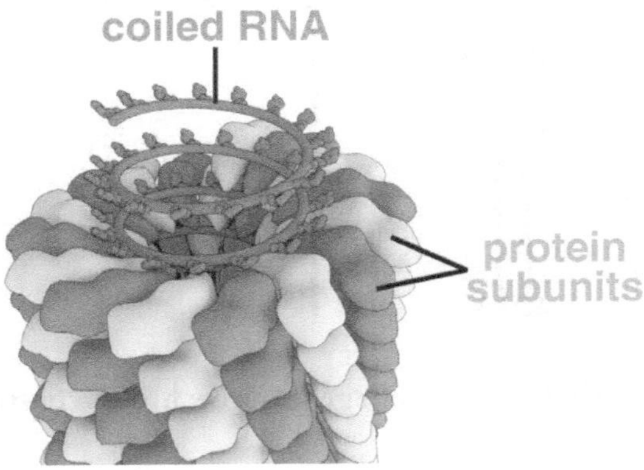

Fig 9.1: Fine structure of a virion Ex. Tobacco Mosaic Virus

The complete complex of nucleic acid and protein packed as a virus particle is called a nucleocapsid.

Some viruses are covered with lipid bilayers, which are referred to as envelopes. Such viruses are referred to as enveloped viruses and those without envelopes as naked viruses.

Virus Symmetry

Symmetry refers to the way in which protein subunits are arranged in the virus coat. The rod-shaped viruses are called helical symmetry, and the spherical-shaped viruses are with icosahedral symmetry.

Ex. For helical symmetry – Tobacco mosaic virus; Icosahedral symmetry – adeno virus.

Envelopes

Complex membrane structures surrounding the nucleocapsid are referred to as envelopes. Enveloped viruses are common in animal viruses (HIV is an enveloped virus). Usually, the envelopes are of lipoprotein nature. The lipid portion is derived from the host cell, and the protein portion is synthesized by the virus. They provide the shape of the virus.

Complex Viruses

Some groups of viruses have complex structures with several shapes. They provide head, tail, end plates, and tail fibers. The maximum of 20 different proteins will be present in these viruses.

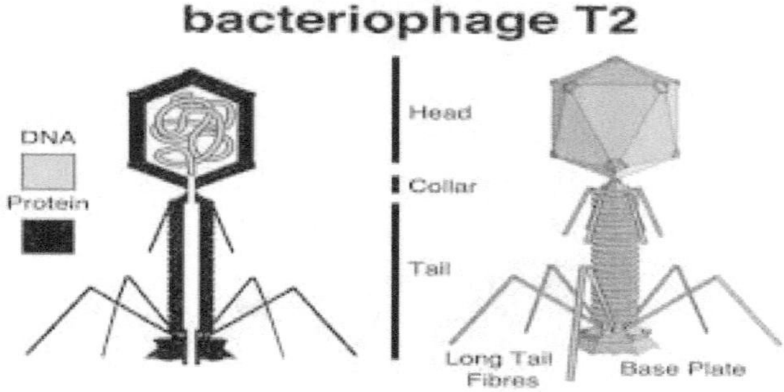

Fig 9.2: Structure of T2 Bacteriophage (Source: https://qsstudy.com/t2-phage-virus/)

Viral Enzymes

As extracelluar particle, virions are metabolically inert. When infect the host cell, they can able to produce some enzymes which are essential for their replication.

1. Nucleic acid polymerase: Either DNA polymerase or RNA polymerase depends upon their genome.
2. Reverse transcriptase: Some RNA viruses (Ex. HIV like retrovirus) have this enzyme to produce DNA from RNA.
3. Neuraminidase: To break down the glycosidic bonds of host cell to enter and to rapture the host cell.
4. Lysozyme: Especially bacteriophages produce lysozyme to make a hole in the bacterial cell wall for entry.

Virus Replication

Since a virus doesn't have any system for replication by itself, it should induce the host cell to synthesize all essential components needed for a complete virus. Then, these components must assemble into proper structure, and new virions must escape from the cell and infect the other cells. The sequence of virus replication is as follows:

1. Attachment (adsorption) of virion to a susceptible host cell
2. Penetration (injection) of virion or its nucleic acid into cell

Early steps in replication: during which the host cell biosynthetic machinery is altered by virion as a prelude to virus nucleic acid synthesis. Virus-specific enzymes are typically made.

1. Replication of virus nucleic acid
2. Synthesis of protein sub-units
3. Assembly (Packing) and Escape or release

Virus Replication Cycle in Host Cell

Attachment

Attachment of virus particles with host cells is mainly based on the specific interaction between host and virus. The virus has one or two specific proteins on the outside of the virion with specific activity. This protein will recognize some host cell surface components such as proteins, polysaccharides, lipoproteins, etc. These compounds are referred to as receptors. If the receptor is removed in the host cell, the attachment of the virus is not possible on the host cell.

Ex. Some bacterial flagella or pili act as receptors; for influenza virus, glycoprotein of red blood cells acts as a receptor; for T2 bacteriophage, core polysaccharide of the outer membrane of E. coli acts as a receptor. The resistant host will have no receptor for the virus, or the receptor is protected by some physical means (e.g., capsules of bacteria give resistance by blocking the receptors).

Penetration

The penetration of the virus into the host cell is mainly based on the surface morphology of the host cell. The penetration varies from host to host (plants and bacteria have cell walls, and animals don't have cell walls).

The bacteriophage (T_2 phage) first attached its tail fibers to the receptor (core protein). By means of lysozyme, the phage made a hole in the cell wall and injected the nucleic acid with the help of the tail. The animal viruses, as a whole virion, penetrate into the host cell through phagocytosis. Plant viruses enter into host cells by mechanical injuries or by some insect's action.

Virus Restriction & Modification by Host

When the viral nucleic acid enters the cytoplasm of the host cell, as a protective mechanism, the host cell can be able to digest the nucleic acid by means of restriction endonuclease (for DNA) or by RNAse (in case of RNA). To avoid that, the viral nucleic acid or genome modifies its base pairs so that the restriction enzymes cannot react to it. By addition of glucose (glucosylation) or by addition of a methyl group (methylation), the modification of the viral genome is possible.

Replication of Viral Nucleic Acid and Proteins

New viral proteins and nucleic acids should be synthesized with the help of host DNA. The genetic system of the host is used by the viral nucleic acid to produce capsid and nucleic acid. Nucleic acids were synthesized by means of DNA or RNA replication using the enzymes DNA polymerase or RNA polymerase, respectively. The protein part of virions is synthesized as protein synthesis. The genetic information of a virus (DNA or RNA) should be converted into mRNA (called transcription). The mRNA synthesis shows diversity among the viruses.

After mRNA is synthesized, viral protein subunits are synthesized. The early proteins—synthesized soon after infection (enzymes)—and late proteins—synthesized later part (coat proteins).

The latent period refers to the period in which the presence of the virus in the host cell cannot be detected. During that period, the synthesis of enzymes, nucleic acids, and proteins takes place.

After the protein coat and nucleic acid synthesis, the assembly takes place leading to the copletion of the viral particle, which will lyse the host cell, escape, and infect the next host cell. The population of virus is expressed as plaque-forming units.

These types of viruses, which kill the host cell during release, are referred to as lytic viruses, and the cycle is referred to as the lytic cycle. But some bacteriophages will multiply inside the host cell and release to the environment after assembly without killing the host cell; they are referred to as temperate phages, and the cycle is referred to as the lysogenic cycle.

1. Bacteriophages

The viruses that attack the bacteria and multiply are referred to as bacteriophages. Most of the bacteriophages studied so far belonged to the viruses affecting E. coli and Salmonella.

- Most of the bacteriophages are naked
- Most of them structurally complex nature
- Head, tail, tail fibres, base plates are some parts of these viruses
- Most of these viruses are used as molecular tool for genetic engineering
- The shape, size, chemistry and replication method are diversified

» ssRNA, dsRNA, ssDNA and dsDNA are as genomes

eplication methods of bacteriophages

1. MS2: This single strand RNA phage uses it RNA as mRNA and protein synthesized. The complimentary RNA also synthesized from which viral RNAs synthesized.
2. X174: This ssDNA phage has single strand circular DNA which duplicate by rolling circle replication.
3. M13: This ssDNA phage has linear DNA. The DNA is synthesized in the cytoplasm, and protein coats are synthesized and accumulated in the cytoplasmic membrane. The linear DNA, when released from the host, was covered by a protein coat and released. This is a temperate phage that will not kill the host cell during replication. (A single decline in the growth of bacteria will occur). The cycle is referred to as the lysogenic cycle, and the phage is a temperate phage, and the host bacterium is a lysogen.
4. Based on the nucleic acid, strands, shapes, the bacteriophages are grouped as follows:
5. T2, T4, T3, T7 phages: They are dsDNA phages which are lytic phages and the cycle is lytic cycle. (25 minutes per cycle). The lytic cycle of these phages occurs as discussed earlier.
6. Lambda phage: This dsDNA phage is a temperate phage. After penetration, the lambda phage DNA will integrate with the bacterial DNA and will persist forever. This state of phage is referred to as a prophage. When the prophage is induced (either by chemical or physical means), it becomes a lytic phage and replicates in the host cell, and by killing the host, the virions will escape.
7. Mu Phage: This dsDNA phage is also a temperate phage like lambda phage. It has the unusual property like transposable elements (which can move from one portion of genomic DNA to another part) and thereby cause mutation. They are referred as biological mutagens.

. Animal viruses

» The animal viruses are both DNA and RNA viruses. Retro virus is a RNA virus but can have DNA as an intermediate (referred as RNA – DNA virus).

» The animal viruses enter into host cell as whole virion through phagocytosis.

» The viral nucleic acid will be produced at the nucleus of host cell and the protein coat will be produced at cytoplasm and assembly takes place.

- Most of the animal viruses are enveloped.
- The animal viruses vary their infection pattern on their host.

Types of infection

- Lytic infection: destroys host cells.
- Persistent infection: host cell continues to shed virus over long time. Ce gradually becomes recognizably poorer (recognized as cytopathic effect, CPE), eventually "crumps out".
- Oncagenes: Infection by certain viruses causes cells to change, becon cancerous. Responsible genes are called oncogenes (tumor-producing genes Viral oncogenes have also been found in uninfected cells. These are gen involved in regulation of cell cycle; when defective, normal regulatory contr is lost and cell can become cancerous.
- Persistent infection: The viruses released slowly from the host cell withou affecting the host cell.
- Latent infection: virus genes may not be expressed for long time (Ex. Man Herpes infections). Not the same as lysogeny – genes are not integrate into host chromosome.

The animal viruses are grouped in to many families based on nucleic acid envelopes, etc.

Retrovirus Replication

The structure of Retrovirus (Human Immunodeficiency Virus) which has ssRN as its genome and surrounded by core protein and envelope.

The retrovirus is the RNA containing enveloped virus, has the specialized syster of replication in the host cells. The virus temporarily synthesizes DNA by means c reverse transcriptase enzyme and the DNA get integrated into:

S. No.	Groups	Enveloped / naked	DNA/ RNA	Imp. Example
1	Parvovirus	N	DNA	Mice minute virus
2	Papovirus	N	DNA	
3	Adenovirus	N	DNA	
4	Iridovirus	N	DNA	
5	Hepadnavirus	E	DNA	Hepatitis B virus
6	Pos virus	E	DNA	Small pox virus
7	Herpes virus	E	DNA	Chicken pox
8	Oucinavirus	N	RNA	Polio virus, Hepatitis A virus
9	Recovirus	N	RNA	Blue tongue virus
10	Togavirus	E	RNA	
11	Rhabdovirus	E	RNA	Rabies virus
12	Orthomyxovirus	E	RNA	Influenza virus
13	Coronavirus	E	RNA	SARS virus
14	Bunyavirus	E	RNA	
15	Retrovirus	E	RNA	HIV
16	Paramyxovirus	E	RNA	Measles virus
17	Arenovirus	E	RNA	

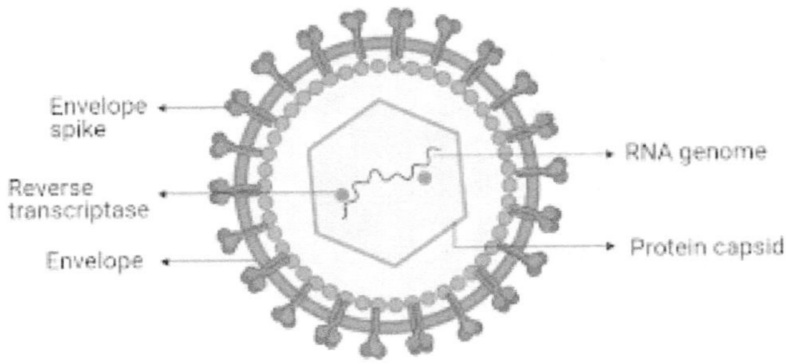

Fig 9.3: Structure of Retrovirus (Ex. HIV) (Sorce: https://www.ahmadcoaching.com/2021/03/how-do-retroviruses-replicate-HIV.html**)**

Fig 9.4: Some animal viruses (Sorce: https://www.sciencedirect.com/topics/agricultural-and-biological-sciences/animal-viruses)

Chromosome of host DNA. After some period, the DNA integrated in the genome synthesizes the RNA and protein coat. After assembly, while escaping from the cell, he cytoplasmic membrane of the host cell provides the envelope, and the virion scaped the host by lysis.

Replication of retrovirus (HIV)

3. Plant viruses

- Most of the plant viruses are naked viruses.
- They don't have a specific mechanism for entering the host plants. The injuries, insect vectors are the means of entry.
- The cell wall and cuticle layers of a plant are the barriers to the entry of viruses.
- The virus diseases on plants are very rare because infection is not strong enough to kill the plant.
- Mostly the virus infection can be recognized by a mosaic-like leaf pattern, in which the loss of chlorophyll leads to yellow in color.
- Some viruses multiply in the host plant without showing the symptom referred to as a latent infection.

There are four important groups of plant viruses.

S. No.	Virus group	Common virus
1	Single strand DNA virus	Tobacco mosaic virus Turnip crinkle virus Cucumber mosaic virus
2	Double strand RNA virus	Wound tumor virus
3	Single strand DNA virus They are referred as Gemini virus (Usually found as pair)	Maize streak virus Tomato golden mosaic virus
4	Double strand DNA virus	Cauliflower Mosaic virus (CaMV) (This virus can modify the plant genome. So used as molecular tool in the field of plant biotechnology

4. Viroids

Viroids are the smallest infectious agents that have a circular RNA alone as a genome (they don't even have a protein coat too). They use DNA-dependent RNA polymerase of the host plant for their multiplication. The viroids infect the plant and cause disease. Ex. Potato spindle tuber viroid. (It consists of 359 ribonucleotide and is characterized by numerous intramolecular base-pairings that lend stabilit to the structure.).

5. Prions

Prions are infectious protein particles that cause several diseases in animals and humans. Prions are the only infectious agents that do not contain any nucleic acids

Prion diseases are often called spongiform encephalopathies because of the postmortem appearance of the brain with large vacuoles. They cause brain disease in sheep (scrapie disease), cows ("mad cow disease"), humans (Kuru and Creutzfeld-Jakob disease), and other animals.

Creutzfeldt-Jakob Disease

- Occurs worldwide and usually becomes evident as dementia.
- Very rare: affects one person in a million. 10 to 15 percent of cases are inherited.
- Typically begins in 60's with a loss of memory.
- Over several weeks the mental deterioration progresses to dementia abnormalities of vision or coordination, rigidity, and involuntary.
- Death usually occurs within six months, and at necropsy the brain shows a "spongy" texture.

Mad Cow Disease

- Cows become uncoordinated and unusually apprehensive.
- Disease traced to use of sheep tissue as food supplement for cows over long period of time in United Kingdom. Practice was begun in late 1970s.

Chapter - 10

Metabolisms of Microbes

he essential nutrients and their function in microbial cells

Element	(%) of dry weight	Source	Some Laboratory reagents used in media preparation	Function
Major elements				
Carbon	50	Organic compounds or CO_2	Glucose, sucrose, lactose, malic acid, mannitol, starch, dextrin	Main constituent of cellular material
Oxygen	20	H_2O, organic compounds, CO_2 and O_2	Water, air and organic compounds	Constituent of cell material and cell water; O2 is electron acceptor in aerobic respiration
Hydrogen	8	H_2O, organic compounds, H_2	Water, air and organic compounds	Main constituent of organic compounds and cell water
Nitrogen	14	NH_3, NO_3, organic compounds, N_2	Ammonium sulphate, sodium sulphate, sodium nitrate, sodium nitrite, peptone, urea, tryptone, soybean extract, beef extract, yeast extract	Constituent of amino acids, nucleic acids nucleic acids nucleotides and coenzymes
Phosphorus	3	Inorganic phosphates (PO_4)	KH_2PO_4, K_2HPO_4, NaH_2PO_4, Na_2HPO_4	Constituent of nucleic acids, nucleotides, phospholipids, LPS, teichoic acids

Sulfur	1	SO_4, N_2S, So, organic sulfur compounds	Na_2SO_4, $MgSO_4$, $Na_2S_2O_3$, cystine like aminoacids	Constituent of cysteine, methionine glutathione, several coenzymes
Potassium	1	Magnesium salts	KCI, KH_2PO_4, K_2HPO_4	Main cellular inorganic cation and cofactor for certain enzymes
Magenesium	0.5	Magnesium salts	$MgSO_4$, MgCI	Inorganic cellular cation, cofactor for certain enzymatic reactions
Calcium	0.5	Calcium salts	$CaCI_2$	Inorganic cellular cation, cofactor for certain enzymes and a component of endospores
Iron	0.2	Iron salts	$FeSO_4$, Fe- EDTA, $FeCI_3$, Fe citrate	Component of cytochromes and certain nonheme iron – proteins and a cofactor for some enzymatic reactions
Micronutrients				
Cobalt	Trace	Inorganic salts	Cobalt chloride	Vitamin B12, cofactor
Copper	Trace	Inorganic salts	Copper sulphate	Respiration, ETC, cofactor
Manganese	Trace	Inorganic salts	$MnCI_2$	Activator of many enzymes
Molybdenum	Trace	Inorganic salts	Sodium molybdate	Flavin containing enzymes, composition in important enzymes
Zinc	Trace	Inorganic salts	Zinc chloride, zinc sulphate	DNA, RNA polymerase enzymes, DNA binding proteins

Apart from these macro- and micronutrients, the microorganisms may require a small amount of certain organic compounds for growth because of their essentiality and inability to synthesize them from available nutrients.

Common vitamins required in the nutrition of certain bacteria

Vitamin	Coenzyme form	Function
p-Aminobenzoic acid (PABA)	-	Precursor for the biosynthesis of folic acid
Folic acid	Tetrahydrofolate	Transfer of one-carbon unit and required for synthesis of thymine, purine bases, serine, methionine and pantothenate
Biotin	Biotin	Biosynthetic reactions that require CO_2 fixation
Lipoic acid	Lipoamide	Transfer of acyl groups in oxidation of keto acids
Mercaptoethane-sulfonic acid	Coenzyme M	CH_4 production by methanogens
Nicotinic acid	NAD (nicotinamide adenine dinucleotide) and NADP	Electron carrier in dehydrogenation reactions
Panotothenic acid	Coenzyme A and the Acyl Carrier Protein (ACP)	Oxidation of keto acids and acyl group carrier in metabolism
Pyridoxine (B6)	Pyridoxal phosphate	Trabsamination, deamination, decarboxylation and racemation of amino acids
Riboflavin (B2)	FMN (flavin mononucleotide) and FAD (flavin adenine dinucleotide)	Oxidoreduction reactions
Thiamine (B1)	Thiamine pyrophosphate (TPP)	Decarboxylation of keto acids and transaminase reactions
Vitamin (B12)	Cobalamine coupled to adenine nucleoside	Transfer of methyl groups
Vitamin K	Quinines and napthoquinones	Electron transport processes

Microbial Metabolism – Basis

Before discussing the microbial metabolism, it is essential to understand the energy flow, basic principles of energetic of living cells etc.

Energy

Defined as the ability to do work. The energy can be referred as Kcal or K Joules. One k cal is the energy required to raise 10C of 1 kg of water. 1 calorie = 4.184 Joules.

Chemical energy – referred as the energy released during oxidation of organic or inorganic compounds.

First Law of Thermodynamics

"Energy can neither be created nor be destroyed in the universe"

So, the living things should collect the energy from existing sources and convert them to a suitable form for the biological process.

Plants grab the energy from light and convert them to high-energ chemical compounds. Similarly, animals can derive their energy by oxidizing the chemicals. Microorganisms, especially the bacteria, have both the ways (from chemicals and sunlight).

Second Law of Thermodynamics

"In all the reactions, some of the energy involved loses it ability to do work"

We can refer this energy lose as "disorder". Generally living systems are "ordered", but universe always prefers disordered. The living systems are always against the disorder and the living systems have constant battle against disorder and when they lose the battle, they die.

Free energy (ΔG)

The quantity of energy which is unable to do work in a reaction is referred as entropy. The energy released from a reaction, that is available to do some useful work is referred as free energy.

Total energy = free energy (AG) + entropy

A reaction may release or require free energy. If a reaction needs energy, it is referred as endergonic reaction and if a reaction releases energy, it is referred as exergonic reaction

A + B – C + D + energy [exergonic reaction]

A + B + energy – C + D [endergonic reaction]

AG of some compounds are

H_2O = -237.2; CO_2 = -396.4; H_2 = 0; O_2 = 0;

N_2O = +104.2; Glucose = -917.3

Negative AG is good and available to do work in the cell.

Activation energy

Referred as the energy required to bring all the molecules to reactive stage.

When the H_2 and O_2 gases kept together in a container will not react each other even after many years and need some energy to make them as reactive molecules to react and gives H_2O. That energy is referred as activation energy.

Catalyst is a substance to lower the activation energy requirement of a reaction and also to increase the reaction rate.

In biological systems, proteins involved in the catalytic activity are referred as biocatalysts or enzymes.

They increase the reaction rate of 10^8 to 10^{20} times.

Enzymes

They are protein in nature with the size of 1 x 10^4 to 1 x 10^6 Dalton.

Normally some non- proteinecious low molecular substances were joint together with protein part to serve as enzymes.

Protein + Non-protein compounds – enzyme

The protein part is Apoenzyme; The non- protein is referred as Co-enzyme or Prosthetic group.

Both (apoenzyme and ecenzyme/Prosthetic group) are referred as Holoenzymes.

The difference between co-enzyme and prosthetic group is that co-enzyme will have loose bound with apoenzyme (temporarily) whereas the prosthetic group will have tight bound with apoenzymes (permanently).

Ex. Co-enzymes – NADP, NAD, FAD, FAM

Ex. Prosthetic group – Heme group of cytochromes

Apart from the co-enzymes and prosthetic groups, the inorganic ions like Ca, Mg, Fe, Mn are required for the activation of enzyme and they are referred as co-factors.

Enzymes are very specific, efficient and can be inhibited by products.

A – Enzymes – C + D

In this reaction, A is the substrate and C and D are the products. For this reaction, the substrate A should attach with the enzyme at specific site referred as substrate binding site. Then, they form a substrate – enzyme complex and lead to act on A to give C and D products. Then, C and D were detached from the enzyme and the enzyme return to its original position.

Function model of an enzyme

Microbial cells are as bags of enzymes. Each enzyme is located at appropriate place and active form. One enzyme converts a substrate to a product and that product may be served as substrate to another enzyme to form product – 2. Like this, series of reactions will form a path way which is referred as biochemical path ways or metabolic path ways.

During this chain of reactions, some reaction may be exergonic and some may be endergonic. So, there is a chance for release of energy or chance for utilization of energy in these reactions.

The released energy from the exergonic reactions were stored in the form of high energy phosphate bonds in ATP (Adenosine triphosphates).

Similarly, whenever the endergonic reactions occur, the ATP break down its high energy bonds and thus released energy can be utilized.

So, the ATP can be referred as energy carrier in the biochemical pathways.

)xidation – Reduction reaction

ı universe, oxidation – reduction reaction in the common method to extract the nergy by living cells.

- All the molecules contain electron as part of atoms.
- These electrons can be disturbed.
- Each molecule can able to donate or accept the electron.

If a molecule or compound donates or loses its electrons, the reaction is referred s oxidation and if a molecule or compound receives or accepts electron, the reaction ; referred as reduction.

Iote: Remember the phrase "LEO says GER" LEO – Loss of Electron is Oxidation; ;ER – Gain of Electron is Reduction

'or example,

If H_2 gas loses its electron,

H_2---------------------------e^- + H^+ will be the reaction,

The H_2 gas will give electron and proton (H^+) as the products. In this reaction, he H_2 loses its electron and hence referred as reduction. But, the reaction is not yet ompleted. If O_2 is present, it will accept the electrons and form water as product.

$2O_2$ + e^- + H^+-----------------------------$2H_2O$.

Since the oxygen received or accepted the electrons, the reaction is referred to s reduction. Here, the hydrogen donates or loses the electron and is hence referred o as the electron donor, and the oxygen receives or accepts the electron and is hence eferred to as the electron acceptor.

Substrates or compounds vary in their ability to donate or accept the electrons in other words, to become oxidized or reduced). The tendency of a substrate to lonate or accept the electron is referred to as oxidation–reduction potential or redox ›otential. The redox potential of each substance was calculated by means of electrical 'olt with reference substance as H2.

The redox potential is negative, meaning the substance is rich in electrons meaning reduced form), and positive means it already lost the electrons (meaning ›xidized form). The reduced forms can give or donate electrons, and oxidized forms can accept the electrons.

If a series of oxidation–reduction reactions occurs, the electrons from or reaction can be carried out to another by means of electron carriers. They can accep the electrons from a reaction and can donate the same in some other reactions. The are called electron carriers. In biological systems, many such carriers are availabl Ex. AD, NADP, FAD, FAM, etc.

In the above reaction, when substrate X gets oxidized, the electron release can be taken up by NADP (oxidized form) and reduced to NADPH. The NADPH (reduced form) then moves to some other reaction and can donate the electron substrate A, which may give the products A and B.

This is how the electron carriers functioned in oxidation-reduction reactio or in metabolic pathways.

Phosphorylation

The addition of phosphorus is referred to as phosphorylation. The energy release during oxidation reactions is stored in the form of high-energy phosphate bonds i ATP. When energy is released, the ADP and Pi react and form ATP. This proce of the addition of Pi to form ATP is referred to as phosphorylation.

Note: The high energy phosphate bonds are differ from normal phosphate bon and the energy released from ATP is higher than glucose phosphates.

The phosphorylation or ATP formation occurs in three different ways in bacteri

a. **Substrate level phosphorylation:** Energy released during oxidation of one organ molecule to another will be used for ATP production. There won't be any extern electron acceptor, and the organic substrate itself acts as the electron accepto This kind of phosphorylation is referred to as substrate-level phosphorylatio
b. **Oxidative or Electron transport phosphorylation:** ATP formation durin electron transport chain by proton motive force is referred as oxidativ phosphorylation. The electrons are carried by several electron carriers and finall the terminal electron acceptor (like O2) will accept the electron and form H2C The proton motive force of ATPase used for synthesis.
c. **Photophosphorylation:** The ATP formation using light energy is referred a photophosphorylation. The proton motive force during photosynthesis is use for ATP synthesis.

Microbial Metabolism

Metabolism refers to the sum of biochemical reactions required for energy generation and the use of energy to synthesize cellular materials.

The energy generation component is referred to as catabolism, and the build-up of macromolecules and cell organelles is referred to as anabolism.

During catabolism, the energy is changed from one compound to another and finally conserved as high-energy bonds of ATP. ATP is the universal currency for energy. When energy is required for anabolism, it may be sent as high-energy bonds of ATP, which has the value of 8 kcal per mole.

Based on the source of carbon, the microbes can be divided into two groups, namely, autotrophs and heterotrophs. Autotrophs utilize CO2 as their sole carbon source, and heterotrophs use organic carbon as their sole carbon source.

Energy generation by heterotrophs

Heterotrophs use a variety of carbon sources. Glucose is the simple and wide variety of microbes that prefer it. The glucose can be taken up by the bacterium through diffusion and can be readily utilized. There are three possible pathways available in bacteria to use glucose.

a. Embden-Meyerhof pathway

b. Phosphoketolase pathway

c. Entner – Doudoroff pathway

All these path ways are fermentative type and substrate level phosphorylation occurs.

Substrate Level Phosphorylation (Fermentation)

The EMP pathway, phosphoketolase pathway, and ED pathway end with one or two ATP syntheses by substrate-level phosphorylation. There won't be any external source of electron acceptor that will come in these reactions.

A. Enbden-Meyerhof pathway

This is the pathway of glycolysis most familiar and common to most of the organisms. The pathway is operated by yeast to produce alcohol and lactic acid bacteria to produce lactic acid and several organic acids, gases, fatty acids, and alcohols.

The pathway is as follows:

The overall reaction is

Glucose – 2 pyruvate + 2 ATP + 2 NADH2

After pyruvate is formed, if the organism is a rrespirative type, the pyruvate will go to Krebs cycle and if the organism is fermentative, the reduction process ends with organic acids, alcohols etc.

A model fermentation: After an intermediate product, a reduction takes place in fermentation, whereas, if respiration, CO2 will be formed by complete oxidation through Krebs's cycle.

Note: After pyruvate, the reduction process leads to fermentation and complete oxidation leads to respiration.

The Embden Meyorhof pathway can lead to a wide array of end products depending on the pathways taken in the reductive steps after the pyruvate formation

The following are some of such fermentations:

Fermentation	**End products**	**Model organism**
Homolactic fermentation	Lactic acid	Lactobacillus
Mixed acid fermentation	Lactate, acetate, formate, succinate	Enterobater
Butyric acid fermentation	Butric acid, acetone	Clostridium acetobutylicum
Propionic acid fermentation	Propionic acid	Propionibacterium
Alcohol fermentation	Ethanol	Saccharomyces

B. Phosphoketolase pathway (Heterolactic pathway)

The phosphoketolase pathway is distinguished by the key cleavage enzyme phosphoketolase, which cleaves pentose to glyceroldehyde 3 phosphate and acetyl phosphate. The pathway ends with ethanol and lactic acid. Ex. Lactobacillus, Leuconostoc. The overall reaction is

Glucose – 1 Iactate + 1 ethanol + 1 CO_2 + 1 ATP

This pathway is useful in the dairy industry for preparation of kefir (fermented milk), yogurt, etc.

The pathway is as follows:

C. Entner – Doudoroff pathway

Only few bacteria like, Zymomonas mobilis employ the ED pathway. The path way is as follows:

The overall reaction is

Glucose – 2 ethanol + 2 CO_2 + 1 ATP

The alcohol productivity of Zymomonas is higher than yeast because of this fermentative pathway.

Note: All the three pathways are end with 1 or 2 ATP by substrate level phosphorylation by means fermentation.

Oxidative Phosphorylation (Respiration)

If the organism is a respiratory type (that means complete oxidation of glucose), it needs following essential metabolic components for their respiration and oxidative phosphorylation.

a. Tricarboxylic acid cycle (also known as citric acid cycle or Kreb's cycle): The pyruvate formed during glycolysis will be completely oxidized to 3 CO2 by the use of this cycle. The metabolic pathway is as follows:

   During oxidation of one pyruvate through TCA cycle, 4 NADH2 1 FADH2 and 1 GTP are produced along with 3 CO2.

b. A membrane associated Electron Transport System (ETC): The electron transport chain is a sequence of electron carriers transport the electrons to a terminal electron acceptor. During this flow of electron in the membrane, a proton motive force across the membrane leads to formation ATP (is referred as electron transport phosphorylation).

c. An outside electron carrier: for aerobic respiration, O2 is the terminal electron acceptor and reduced to H2O. This is normal for higher organisms. But in anaerobic bacteria, the terminal electron acceptor may be of nitrite, nitrate or carbon dioxide.

In aerobic organisms, the terminal electron acceptor will be O2. In some anaerobic organisms, after the electron transport chain, instead of O2, some inorganic compounds like sulphate, nitrate or some organic compounds like fumarate act as terminal electron acceptor. Such type of respiration is referred as anaerobic respiration and the normal O2 mediated respiration is referred as aerobic respiration.

Energy generation by autotrophs

Autotrophs use CO_2 as their sole carbon source. There are two types such as photoautotrophs and chemoautotrophs.

Photoautotrophs use light as energy source and CO2 as carbon source.

Chemoautotrops use chemicals (especially inorganic) as energy source and CO2 as carbon source.

Energy and carbon assimilation by photoautotrophs

Phototrophs use sunlight to produce ATP through phosphorylation, referred as photophosphorylation. The photorophs convert the light energy to chemical energy (ATP) through the process called photosynthesis.

Photosynthesis is a type of metabolism in which catabolism and anabolism occur as sequence. The catabolic reaction (energy generating process) of photosynthesis is light reaction in which the light energy is converted to chemical energy (ATP) and electrons or reducing powers (NADPH). The anabolic reaction (macromolecule synthesis) of photosynthesis is dark reaction in which CO_2 is converted to organic molecules (carbohydrates), which is also called as CO_2 fixation.

For conversion of light energy to ATP, the bacteria possess light harvesting pigments. They are chlorophyll a, carotenoids, phycobiliproteins (which are present in cyanobacteria) and bacteriochlorophyll (which are present in purple sulphur bacteria).

In bacteria, there are two types of light reactions (conversion of light to ATP) and two types of CO_2 fixation occur.

Photophosphorylation

For photophosphorylation, a light harvesting pigments, a membrane electron transport chain, source of electron (electron donor) and ATPase enzymes are required.

Two types of photophosphorylations occur during photosynthesis. Cyclic hotophosphorylation and non-cyclic photophosphorylation.

» In plant and cyanobacteria, both cyclic and non-cyclic photophosphorylation occur whereas in purple bacteria, the cyclic photophosphorylation only occurs.

» In plant and cyanobacteria, the electron source is water, by photolysis, H_2O split into H^+ and O_2 and during the process, O_2 is evolved and referred as oxygenic photosynthesis.

» Since, the sulphur bacteria is an anaerobic bacterium, they use H_2S instead of H_2O as electron donor. Since, there won't by any O_2 evolution during photosynthesis, referred as anoxygenic photosynthesis.

)ifference between plant and bacterial photosynthesis

Organisms	**Plant photosynthesis**	**Bacterial photosynthesis**
Type of chlorophyll	Plants, algae, cyanobacteria	Purple and green bacteria
	Chlorophyll a absorbs 650-750 nm	Bacteriochlorophyll absorbs 800-1000 nm
Photosystem I (cyclic photophosphorylation)	Present	Present
Photosystem II (noncylic photophosphorylation)	Present	Absent
Produces O_2	Yes (Oxygenic)	No (Anoxygenic)
Photosynthetic electron donor	H_2O	H_2S, other sulfur compounds or certain organic compounds

The oxygenic photophosphorylation is as follows:

The end product of the light reaction is ATP, NADPH and O_2. The ATP and NADPH, the energy and electron sources thus produced were used for dark reaction.

The anoxygenic photo phosphorylation will take place as in the image and the end product of the light reaction is ATP, NADPH and Sulphur. The ATP and NADPH, the energy and electron sources thus produced were used for dark reaction.

Dark reaction (CO_2 fixation)

The dark reaction in which the ATP and NADPH were used as energy and electro sources to fix the CO_2 as carbodhydrates. The pathway involved in the dark reaction calvin cycle, by which the CO_2 is fixed as phosphoglyceic acid and lead to formatic of many sugars. The enzyme RuBiSCO is the key enzyme forthis process.

The following pathway shows the Calvin cycle and the formation of ke monomers for anabolic reactions such as hexose phosphate – polysaccharides; pyruv acid – amino acid and fatty acid, pentose phosphate – DNA and RNA.

Another way of CO_2 fixation by phototrophs

In phototrophs, the electron and energy were derived form sunlight and carbo from CO_2 fixation through Calvin cycle. But some bacteria may derive electro and energy from sunlight and fixes CO_2 by some other path way not, Calvin cycl The example is Photosynthetic green bacteria (Chlorobium). They derive NADPI and ATP through cyclic phosphorylation, but CO_2 fixation is by reverse TCA cycl Since TCA cycle is amphibolic pathway (referring the cycle can operate in both th directions), it can also be used to fix the carbon-di-oxide if operated reversely. Th pathway is as follows:

Another way of CO_2 fixation is by methanogens. They use CO_2 as termin electron acceptor and forms CH_4 (methane). They also fix by acetyl CoA pathwa for fixing CO_2.

A complete model of light and dark reaction of photosynthesis

Energy and carbon assimilation by Chemoautotrophs

Since the chemoautotrophs use inorganic chemicals for their energy and electro source, they are referred as chemolithotrophs or chemolithotrophic autotrophs.

These organisms remove electron from an inorganic substance and put then through electron transport chain for ATP synthesis (through electron transpor phosphorylation). At the same time, the electrons also flow through reverse electro transport chain and with the end product of NADPH. These ATP and NADPH are used for CO_2 fixation through Calvin cycle.

These bacteria are obligate aerobic organisms. Some examples of th chemolithotrophs are as follows:

Groups of chemolithotrophs

Physiological group	Energy source	Oxidized end product	Organism
Hydrogen bacteria	H_2	H_2O	Alcaligenes, Pseudomonas
Nitrifying bacteria	NH_2	NO_2	Nitrosomonas
Nitrifying bacteria	NO_2	NO_3	Nitrobacter
Sulfur oxidizing bacteria	H_2S or S	SO_4	Thiobacillus, Sulfolobus
Iron oxidizing bacteria	Fe^{2+}	Fe^{3+}	Gallionella, Thiobacillus

Chapter - 11

Microbial Control

Sterilization refers to the complete removal or elimination of microorganisms from an environment. Physical and chemical agents are available to remove the microorganisms from the environment. Heat, irradiation, and filtration are the three physical agents, and various chemicals are available for sterilization.

Terms related to control of microorganisms

Cide or cidal means to kill the microorganism (Ex. Bactericidal refers the killing of bacteria).

Static or stasis means inactivation or inhibition, but not killing. They prevent multiplication of the organism. (Ex. Fungistatic refers the inhibition of growth of fungi).

Sepsis refers break down of a living tissue by organisms and is accompanied by inflammation and pus formation.

Antispectic is an agent applied externally on living tissues to kill or inhibit the growth of the organisms.

Disinfectant is an agent applied externally on inanimate objects (not living tissues) to destroy harmful pathogens in their vegetative stage.

(Antiseptics are milder than disinfectants because to avoid the side effects)

Sanitation refers removal of organisms from a location by cleaning but not by sterilization.

A. Physical Agents

1. Heat

Principle: At high temperature, all the macor molecules lose their structure and ability to function called as denaturation. Similarly, the heat also coagulate the protein and oxidize the elements lead for lethality.

The heat susceptibility of the organism should know before the heat treatment to be applied. For this, the thermal death time and thermal death points are essential.

Thermal death time refers the time to kill a suspension of organisms at a specific temperature.

Thermal death point refers the lowest temperature to kill a suspension in a given time.

Decimal reduction time (D time) – Time to kill 90% or 1 log unit of a population at a given temperature.

Based on the above parameters, the time and temperature to kill the microorganism from an environment is derived.

4 or 5 different ways of sterilization are available to kill the microorganisms using heat.

1. Direct flaming: Showing the objects or equipments directly to the flame leads the sterilization. This is also called as incineration. The inoculation needles, spread rods and forceps can be sterilized using this technique.
2. Boiling: In this method, only the vegetative form are killed in few minutes whereas the spores are not.
3. Dry heat: Causes oxidation of cells. By using electrical coils, heat is generated and used for sterilization. Normally, high temperature and long time are required for complete sterilization using this method. 160-180°C for 2-3 hours is required for complete sterilization. A hot air oven is an instrument used for dry heat sterilization. It is a double-walled (inner and outer walls) instrument protected with asbestos sheets to avoid heat loss. The thermostat is present to control the temperature, and the timer is also present to turn off automatically after a particular time. The glassware, like conical flasks, beakers, petri dishes, and pipettes, can be sterilized using this technique.
4. Moist heat: causes denaturation and coagulation of protein. To kill vegetative structure of bacteria, yeast, and molds—80°C for 5-10 min.; to kill mold

spores—80°C for 30 min.; bacterial spores—121°C, 15 lbs/sq. inch pressur for 15 min. are the optimum moist heat. The autoclave is used for sterilizatio using moist heat under pressure. An autoclave is made up of double-walle steel plates with airtight lids. An electrical coil submerged in water is provide to generate the stream. To measure the pressure, a pressure gauge and, fo safety, a safety valve are also attached to the lid. If the stream pressure insid the closed vessel is increased to 15 lb/sq. inch, the temperature will raise t 121.6°C. Keeping this condition for 15-20 min. will kill all the vegetativ and spore forms of microbes.

(The moist heat has more penetration power than dry heat and leads faste reduction in number of living organism and under pressure, the penetration of hea will be more in the autoclave).

Tyndallization

Refers to intermittent or fractional sterilization. The subsequent cooling and heating by steam for 3 days will remove the germs and their spores, referred to as tyndallization Ex. Soil that contains diversified microbes with spores. The tyndallization will remove all the vegetative cells and also the spores. The normal sterilization by autoclave wil eliminate the vegetative cells only and not spores

Pasteurization

Refers to the removal of undesired microorganisms without affecting the beneficia microorganisms, odor, and taste. The pasteurization of milk is done to eliminate the pathogenic microorganisms that cause tuberculosis, brucellosis, Q fever, typhoid etc. In broad terms, the pasteurization is done to kill Salmonella and E. coli-like organisms.

Flash pasteurization: holding the substance at 71^{0}C for 15 seconds and rapid cooling referred as flash pasteurization.

Bulk pasteurization: holding the substance at 63-65^{0}C for 30 min and slow cooling referred as bulk pasteurization

Tyndallization > sterilization (by autoclave) > pasteurization is the order of degree of strength in terms of removal of microorganisms.

I. Radiation

Iicrowaves, ultra violet rays, gamma rays, electrons have the power to kill the icroorganisms. Among them, UV light (non-ionizing radiation) and gamma rays onizing radiation) are commonly used for sterilization.

V radiation

he wave length considered to be between 220 – 300 nm is used to control the icroorganisms. It has sufficient energy to form dimmers (bridges) between two djacent pyrimidines in the DNA and thereby affect the DNA replication leads death. This light is highly useful to disinfect surfaces, air, and water, which do ot absorb the light. Laminar air flow chamber – A clean bench used to handle e microbes in a microbe free environment. It has two methods of sterilization. adiation by UV and filtration by HEPA filters (High efficiency particulate filters), microbe free air will be present in side.

onizing radiation

onizing radiation is electromagnetic radiation of sufficient energy to produce ions ke electron, hydroxyl lions, protons, etc. which can alter the biopolymers suc has NA and protein causing break of polymers. γ-rays produced from ^{60}Co and ^{137}Cs re useful for sterilization of the plastic wares, gloves, heat labile plastic containers, oly bags, tissue grafts, plastic syringes etc. In few cases, food were also sterilized y γ-rays.

II. Filtration

Ieat is the most common and effective way of sterilization, some heat sensitive quids cannot be sterilized by heat. An alternate technique for the sterilization is y filters. A filter is too small for a passage of microorganisms but large enough to assage of liquids.

here are three types of filters

a. Pyrex glass or sintered glass filters: Filter glass made with powdered glass disc pore size of 0.5 to 2 nm discs. These discs were fit into funnels and used for filtration. These were used by suction pumps for faster flow rate.

b. Seitz filters: The filter mat is made up of asbestos – cellulose mixture.

c. Membrane filters: They are composed of polymers with high tensile strength such as cellulose acetate, cellulose nitrate or polysulfonate, manufactur in such a way that they contain a large number of tiny holes. By adjusti the polymerization conditions during manufacture, the size of holes in t membrane can be precisely controlled. They trap the microbes on the surfa and allow the microbe free solution to pass.

B. Chemical Agents

Different chemicals like halogens, heavy metals, phenols, alcohols, ethanol are us for sterilization.

The following are different classes of chemicals involved in sterilization.

1. Phenol and Phenolics: 5% aqueous solutions kill vegetative forms. Some their derivatives are Lysol, Dettol, and Cresol etc. They alter the selecti permeability of the cytoplasmic membrane of organisms.
2. Alcohols: They are bactericidal and fungicidal but not sporicidal. Th denature proteins and are solvents of lipids. 60-90% is the range for th activity. Water is required for the lethal effect.
3. Halogens: Cause oxidation and direct halogenations of proteins th inhibiting the activity of proteins e.g. hypochlorite, iodophores etc.
4. Surfactants: Surface active agents, good wetting and solubilizing agents e. Soap, Detergent etc. They have both hydrophilic and hydrophobic grou (ambivalent) to dissolve compounds.
5. Alkylating agents: They substitute alkyl groups for hydrogen of reactive grou in nucleic acids and proteins, causing disruption of metabolic pathways e. Formaldehyde, glutaraldehyde, ethylene oxide (gaseous agent).
6. Heavy metals: reacts with sulfhydryl group (SH) of enzyme and make the inactive. Mercury, copper and silver is commonly used heavy metals.
7. Antibiotics: Antibiotics are low molecular weight molecules produced secondary metabolites by microorganisms, which kill or inhibit the oth microorganisms. The following table shows the summarized examples common antibiotics and their action.

Chemical	Use	Mode of action
Antisepsis		
Alcohol	Skin	Lipid solvent and protein denaturant
Phenol compounds	Soaps, lotions, cosmetics, body deodorants	Disrupt cell membrane
Cationic detergents (Ex. Benzalkonium chloride)	Soap, lotion	Phospholipid interaction
3% hydrogen peroxide	Skin	Oxidizing agent
Lodine containing iodophors	Skin	Lodinates tyrosine residues; oxidizing agent
Silver nitrate	Eyes of new born to prevent blindness from Nesseria gonorrhoea	Protein precipitant
Disinfectants		
Alcohol	Medical instruments, food, dairy instruments, lab surface	Lipid solvent & protein denaturant
Cationic detergents	Medical instruments, food, dairy instruments, lab surface	Interact with phospholipid
Chlorine gas	Disinfectant for purification of water supplies	Oxidizing agent
Chlorine compounds (Chloramines, sodium hypochlorite, chlorine dioxide)	Disinfectant for food and dairy industry, water supplies	Oxidizing agent
Copper sulphate	Algicide in swimming pools	Protein precipitant
Ethylene oxide (gas)	Sterilant for temperature sensitive lab material like plastics	Alkylating agent
Formaldehyde	3-8% used as surface disinfectant; 37% (referred as formalin) as preservatives	Alkylating agent
Gluteraldehyde	2% solution	Alkylating agent
3% hydrogen peroxide	Medical instruments	Oxidizing agent
Mercuric chloride	For laboratory surface, plant samples	Protein denaturation
Ozone	Drinking water	Oxidizing agent

Preservatives		
Salt and sugar solutions	Preservative	Osmotic pressure

S. No.	Antibiotics	Source	Effective against	Mode of action
1	Penicillin	*Pennicillium notatum, P. crysogenum*	Gran +ve bacteria	Inhibit the ell wall (Peptidoglycon layer) synthesis
2	Amphicillin		Gram +ve bacteria	Inhibit the ell wall (Peptidoglycon layer) synthesis
3	Streptomycin	*Streptomyces griseus*	G + and G –ve organisms	Inhibit the translation process of protein synthesis
4	Erythromycin	*Streptomyces crythreus*	G + and G -ve	Inhibit the translation process of protein synthesis
5	Polymixin	*Bacillus polymyxa*	Gram -ve	Damages Cytoplasm
6	Tetracycline	*Streptomyces sp.*	G + and G-ve	Inhibit the translation process of protein synthesis
7	Chloremphe necol	*Streptomyces sp.*	Bacteriostatic	Inhibit the protein synthesis

Chapter - 12

Concepts of Microbial Genetics

Genetics is the study of heredity and variations of genes. How the genes carry information, expressed and replicated in subsequent generations of organisms.

The molecular basis of a gene is deoxyribonucleic acid (DNA), which exists in the nucleus. DNA is made up of a double-stranded helix; the two strands are held together by hydrogen bonds between specific nitrogenous base pairs: A-T and C-G.

A gene is a segment of DNA, a sequence of nucleotides, that codes for a functional product, usually a protein.

When a gene is expressed, DNA is transcribed to produce RNA; mRNA is then translated into proteins. This is also referred to as the Central Dogma of Life. The DNA in a cell is replicated before the cell divides, so each daughter cell receives the same genetic information.

Genotype and Phenotype

Genotype is the genetic composition of an organism its entire DNA. Phenotype is the expression of the genes the proteins of the cell and the properties they confer on the organism.

DNA and Chromosomes

The DNA in a chromosome exists as one long double helix associated with various proteins that regulate genetic activity.

Bacterial DNA is circular; the chromosome of E. coli, for example, contains about 4 million base pairs and is approximately 1000 times longer than the cell.

Genomics is the molecular characterization of genomes. Information containe in the DNA is transcribed into RNA and translated into proteins.

Enzymes that associated with DNA

1. DNA polymerase: An enzyme that synthesize a new strand of DNA i 5' – 3' direction using anti parallel strand as template.
2. Restriction Endonuclease: An enzyme which recognizes and makes doubl strand DNA breaks at specific sequence of DNA.
3. RNA polymerase: An enzyme that synthesize RNA in 5' – 3' direction usin anti parallel DNA strand as template.
4. DNA ligase: An enzyme that lygate or joint double strand DNA fragment
5. DNA gyrase: These enzymes introduce super coiling of DNA in prokaryote (also referred as Topoisomerase II).
6. Topoisomerase I: An enzyme which removes the super coiling of DN (Super coiling refers the highly twisted form of DNA).

DNA Replication

During DNA replication, the two strands of the double helix separate at the replicatio fork, and each strand is used as a template by DNA polymerases to synthesize tw new strands of DNA according to the rules of nitrogenous base pairing.

The result of DNA replication is two new strands of DNA, each having base sequence complementary to one of the original strands. Because each double stranded DNA molecule contains one original and one new strand, the replicatio process is called semiconservative. DNA is synthesized in one chemical directio called 5'–3' (5' is the phosphate end; 3' is the hydroxyl end of deoxyribose). At th replication fork, the leading strand is synthesized continuously and the laggin strand, discontinuously. DNA polymerase proofreads new molecules of DNA an removes mismatched bases before continuing DNA synthesis. Errors only occur time for every 1010 bases added. Each daughter bacterium receives a chromosom identical to the parent's.

Transcription

During transcription, the enzyme RNA polymerase synthesizes a strand of RNA from one strand of double stranded DNA, which serves as a template.

RNA is synthesized from nucleotides containing the bases A, C, G, and U, which ir with the bases of the DNA sense strand. The starting point for transcription, here DNA polymerase binds to DNA, is the promoter site; the region of DNA at is the endpoint of transcription is the terminator site; RNA is synthesized in e 5'–3' direction.

ranslation

ranslation is the process in which the information in the nucleotide base sequence f mRNA is used to dictate the amino acid sequence of a protein. The mRNA ssociates with ribosomes, which consist of rRNA and protein.

Three-base segments of mRNA that specify amino acids are called codons. he genetic code refers to the relationship among the nucleotide base sequence f DNA, the corresponding codons of mRNA, and the amino acids for which the odons code. The genetic code is degenerate; that is, most amino acids are coded r by more than one codon. Of the 64 codons, 61 are sense codons (which code r amino acids), and 3 are nonsense codons (stop codons), which do not code for mino acids and are stop signals for translation. The start codon, AUG, normally odes for methionine (formylmethionine at the beginning of a protein). Specific mino acids are attached to molecules of tRNA. Another portion of the tRNA has base triplet called an anticodon. The base pairing of codon and anticodon at the bosome results in specific amino acids being brought to the site of protein synthesis. he ribosome moves along the mRNA strand as amino acids are joined to form a rowing polypeptide; mRNA is read in the 5'–> 3' direction. Translation ends when he ribosome reaches a stop codon on the mRNA.

acterial Genome Organization

NA molecules that replicate as discrete genetic units in bacteria are called replicons. n some Escherichia coli strains, the chromosome is the only replicon present in he cell. Other bacterial strains have additional replicons, such as plasmids and acteriophage.

hromosomal DNA

Bacterial genomes vary in size from about 0.4 x 109 to 8.6 x 109 daltons (Da), ome of the smallest being obligate parasites (Mycoplasma) and the largest belonging o bacteria capable of complex differentiation such as Myxococcus. The amount of NA in the genome determines the maximum amount of information that it can encode. Most bacteria have a haploid genome, a single chromosome consisting o circular, double-stranded DNA molecule. However, linear chromosomes have be found in Gram-positive Borrelia and Streptomyces spp., and one linear and o circular chromosome are present in the Gram-negative bacterium Agrobacteriu tumefaciens.

The single chromosome of the common intestinal bacterium E. coli is 3 x 1(Da (4,500 kilobase pairs [kbpl] in size), accounting for about 2 to 3 percent of t dry weight of the cell. The E. coli genome is only about 0.1% as large as the hum genome, but it is sufficient to code for several thousand polypeptides of average si (40 kDa or 360 amino acids).

The chromosome of E. coli has a contour length of approximately 1.35 m several hundred times longer than the bacterial cell, but the DNA is supercoiled a tightly packaged in the bacterial nucleoid. The time required for replication of t entire chromosome is about 40 minutes, which is approximately twice the shorte division time for this bacterium. DNA replication must be initiated as often as t cells divide, so in rapidly growing bacteria, a new round of chromosomal replicati begins before an earlier round is completed. At rapid growth rates, there may be fo chromosomes replicating to form eight at the time of cell division, which is couple with the completion of a round of chromosomal replication. Thus, the chromoson in rapidly growing bacteria is replicating at more than one point. The replicatio of chromosomal DNA in bacteria is complex and involves many different protein

Plasmids

Plasmids are replicons that are maintained as discrete, extra chromosomal genet elements in bacteria. They are usually much smaller than the bacterial chromosom varying from less than 5 to more than several hundred kbp, though plasmids as lar as 2 Mbp occur in some bacteria. Plasmids usually encode traits that are not essenti for bacterial viability and replicate independently of the chromosome. Most plasmi are supercoiled, circular, double-stranded DNA molecules, but linear plasmids ha also been demonstrated in Borrelia and Streptomyces.

Many plasmids control medically important properties of pathogenic bacteri including resistance to one or several antibiotics, production of toxins, and synthes of cell surface structures required for adherence or colonization. Plasmids tha determine resistance to antibiotics are often called R plasmids (or R factors), an the plasmids responsible for conjugation are F plasmids.

Gene regulation & expression

The genes are always present in the bacterial chromosomes, but their expression is always controlled. Here the gene expression refers to the mRNA and protein synthesis. The expression of the gene was systematically controlled by some proteins. Whenever needed, the genes will be expressed to get the protein (i.e., enzymes).

Regulating protein synthesis at the gene level is energy efficient because proteins are synthesized only as they are needed. Constitutive enzymes are always present in cells. Examples are genes for most of the enzymes in glycolysis. For these genetic regulatory mechanisms, the control is aimed at mRNA synthesis.

The Operon Model of Gene Expression

In bacteria, a group of coordinately regulated structural genes with related metabolic functions and the promoter and operator sites that control transcription are called an operon. It is also defined as a cluster of genes whose expression is controlled by a single operator. An operon essentially consists of the following components:

1. Promoter: A site on the DNA where the RNA polymerase binds and begin the protein sysnthesis.
2. Operator: A site on the DNA where the repressor protein binds and blocks the mRNA synthesis.
3. Repressor protein: A protein bind in the operator region and block the mRNA synthesis.
4. Structural genes: They are the actual coding region of the proteins or enzymes. This portion of the gene is also referred as open reading frame, because this portion alone is ready by RNA polymerase to transcript the mRNA.
5. Terminator: This portion of DNA of the operon will terminate the mRNA synthesis by forming loop like arrangement in the same strand.
6. Ribosome binding site: Some 6 base pairs next to promoter, where ribosome bind with mRNA during translation process

Lac operon – suppression & Expression

P denotes the promoter region, in which the RNA polymerase binds; O denotes the operator region, in which the suppressor protein binds and blocks.LacZ, LacY, and LacA are the structural genes that code the enzymes, namely β-galactosidase, lactose permease, and lactose transacetylase (which are essential for lactose utilization). T denotes the terminator, which stops the mRNA synthesis.

Under normal conditions, meaning no lactose in the medium, the repressor block the operator of the lac operon, and thereby no production of these enzymes occurs

When lactose was introduced to the medium, it was transported to the insid of the cell and acted as an inducer. The inducer will bind with repressor protei and lead to no more blockage of the operator region. Then RNA polymerase start synthesizing mRNA up to the terminator region. The transcription will happen unti the presence of inducer (lactose). When lactose was exhausted in the medium, th repressor protein will again block the operator, and no more mRNA synthesis wil take place. This kind of gene regulation is referred to as repression, and sometime induction-type gene regulations are also common ones.

Importance of Microbial Genetics

1. To understand the gene functions of microorganisms.
2. Microbes provide relatively simple system for studying genetic phenomenon and thus useful for other higher organism.
3. Microorganisms are used for isolation and multiplication of specific genes of higher organisms (which is referred as gene cloning).
4. Microbes provide many value added products like antibiotics, growth hormones etc. Microbial genetics will be helpful to increase these products productivity by microbial biotechnology.
5. Understanding the genetics of disease causing microorganisms especially virus, will be useful to control the diseases.
6. Gene transfer among the prokaryotes play major role in the spread of the genes in a particular environment. Microbial genetics will be useful to study the gene transfer form one organism to another.

Mutation

Is an inherited change in the base sequence of the nucleic acid comprising the genome of an organism. A strain carrying such a change is called a mutant strain, and a non-mutated strain that was isolated from the natural environment is called a wild-type strain.

The genotype of an organism for a particular character can be written as three-lettered lowercase italic letters following a capital letter. Ex. lacZ refers to the gene that codes the enzyme β-galactosidase. The protein transcribed from this can be written as normal letters starting with a capital letter. For the above gene, the protein may be written as LacZ. The phenotype of the organism for this gene (whether present r not) can be written as LacZ+ for presence and LacZ- for absence.

Mutant identification

n a wild type organism very rarely the mutation will occur with very low mutation requency. (may be 1 to 10^6 to 10^8 cells). So, identifying the mutant from the wild ype is a key phenomenon.

The mutants can be identified from a population either by screening or by election. Screening refers to the procedure that permits the sorting of organisms y phenotype or genotype. Selection refers to the procedure in which pressure will e given to grow a particular genotype.

For example, in a mixed population, a particular mutant lost its color pigmentation); we have to go for a screening procedure to identify the mutant. n this case, we cannot use selection to identify the mutant. This type of mutant dentification is very laborious and limited.

On the other hand, in a mixed population, a particular mutant gains an antibiotic esistance (the wild type may be susceptible); we can use a selection procedure to dentify the mutant by growing the mixed population over the medium incorporated with the particular antibiotic, which will allow the mutant alone to grow on that. This type of mutant identification (selection) is a powerful tool to identify the mutant.

Replica Planting Technique

This is also a selection type of mutant identification procedure that will be useful to identify auxotroph mutants. An auxotrophic mutant is a metabolically defective mutant. This mutant, which cannot synthesize a particular nutrient, required the nutrient externally and is referred to as an auxotrophic mutant. The strains that can synthesize the nutrients by themselves are referred to as prototrophs. The auxotrophic mutants are very common for amino acids. Normally, the wild strains can synthesize all the essential amino acids by themselves. By mutation, if a strain lost its ability to synthesize a particular amino acid (e.g., leucine), the mutant is referred to as a leu auxotroph mutant. This mutant can grow only in the medium supplemented with leucine, whereas the prototroph can grow in the medium without leucine. This type of auxotrophic mutant can be selected by the replica plating technique.

The procedure of replica plating techniques is as follows: The master plate (a mixture of prototrophs and auxotrophs) will be duplicated to medium with and without leucine by means of a rubber stamp-like device (called a replica plate). A clone that grows in medium with leucine and which cannot show growth on mediu without leucine is an auxotroph.

Spontaneous & Induced mutation

The natural radiation (cosmic rays, UV rays etc) alter the genotype of an organism which is called as spontaneous mutation. Induction of mutation by some agent artificially is referred as induced mutation.

Point Mutation

Mutation occurs in one or two pairs of a gene is referred as point mutation. Th other mutation is change in the segment of chromosome, referred as chromosoma aberration. Since bacteria are having single filamentous double strand DNA as thei genome, the point mutation is a common one.

The point mutation occurs by means of two ways as A. Base pair substitutio B. Frame shift mutation.

A. Base Pair substitution

One nucleotide base pair is replaced by another base pair; such a mutation is referre to as a base pair substitution. One purine (adenine or guanine) is replaced by anothe purine, or one pyrimidine (thymine or cytosine) is replaced by another pyrimidine is called a transition. One purine is replaced by a pyrimidine, or one pyrimidine i replaced by a purine, is called a transversion.

The base pair substitution type mutation alters the base pair arrangement of a gene and leads to an alteration in the transcription (mRNA) and translation (protein) There are three types of effects due to base pair substitution.

1. **Silent mutation:** The change of base pair, that codon codes the same aminc acid (as that of the wild type), such a mutation is a silent mutation. (Becaus there are four codons available for a single amino acid). In this, a mutatior occurs, but the amino acid sequence of the protein is not changed.
2. **Nonsense mutation:** The change of base pair, which causes the codon tc stop the protein synthesis (due to the change to a stop codon), the effect is referred to as a nonsense mutation. This base pair substitution will stop the protein synthesis that yields the incomplete protein.
3. **Missense mutation:** The change of base pair, that codon codes some different

amino acid; such an effect is referred to as an nonsense mutation. Here, the protein will be synthesized but with an altered amino acid sequence.

. Frame Shift Mutation

. type of point mutation is which mutation occurs by addition or delection of a ase pair, which is referred as frame shift mutation.

Addition or deletion of base pair leads to change or shift the reading frame uring transcription resulted in change in the amino acid sequence of protein.

Addition of base pair is referred as addition mutation and deletion of base pair deletion mutation.

.eversion

he mutation which discussed above is reversible, that means the mutants can revert) the wild type. This phenomenon is referred as reversion mutation and the strain . called as revertant.

A second site mutation which restores the wild type genotype is referred as ıppression mutation.

nsertion sequences

.bout 700 – 1400 base pair sized DNA fragments (as identified genes) inserted ı the wild genotypes, cause mutation are referred as insertion sequences. The gene esponsible for such character of insertion sequence is transposase gene.

'henotypic variations observed in microbial mutations

he following are some common phenotypic characters observed during mutation.

Phenotypes	Cause
Auxotrophs	Loss of enzymes to biosynthesis a particular nutrier (especially amino acids)
Cold sensitivity	Alter the cold resistant protein
Drug resistance	Increase the permeability and detoxification of drug
Non-capsulate	Loss of capsules
Loss of pigments	Loss of enzymes for biosynthesis of pigments
Rough colony	Loss of polysaccharides particularly extracellular, whic lead to dry and rough unshaped colonies
Sugar fermentations	Loss of fermentative path way enzymes
Non-motile	Loss or unfunctional flagella
Temperature sensitivity	Alter the heat resistance proteins
Virus resistance	Loss of viral receptors

Mutagens

The chemical, physical and biological agents that can induce the mutation are referr as mutagens. The following are some of the features of mutagens.

1. Chemical mutagens

The chemicals can be divided into four groups based on their action namely.

» Base analogs – Incorporate in the DNA instead of particular N base;

» DNA reacting chemicals – react with DNA nucleotide (especially sugar

» Alkylating agents – change the structure of DNA and

» Intercalating dyes – These chemicals will insert between base pairs of DN

S. No.	Chemical	Action on base pair of DNA	Result
a. Base analogs			
1.	5-bromouracil	Incorporated instead of T	AT – GC
2.	2-aminopurine	Incorporated like A	AT – GC
b. Chemicals react with DNA			
1.	Nitrous oxide	Deaminates A and C	AT – GC, GC – AT
2.	Hydroxylamine	Reacts with C	GC – AT

c. Alkylaitng agents			
1.	Ethyl methane sulfonate (EMS)	Add methyl group on G lead to faulty pairing with T	GC – AT
2.	Nitrosoguanidine	Cross link the DNA strands	Delection mutation
d. Intercalating agents			
1.	Acrydine orange, Ethydium bromide	Insert between base pairs	Addition mutation

2. Physical agents

Both non-ionizing and ionizing radiations able to cause mutation. Non-ionizing radiation has wider use in the microbial mutations.

» **UV light:** This non-ionizing radiation can be absorbed by puring and pyrimidine strongly at 260 nm. Two adjacent pyrimidines form a dimer, by which covalently joint. During replication, the DNA polymerase wrongly recognize the pyrimidine and add a wrong base pair which leads to mutation.

» **Gamma rays:** This ionizing radiation at low concentrations react as free radical attack and DNA breakage which leads to mutation. Gamma rays are not commonly used for microbial mutations, instead, highly useful for mutating plant materials.

3. Biologcal agents

Transposable element, insertion sequences and Mu phage are some of the biological agents cause mutation.

» Transposable element: A genetic element that can move from one place of chromosomal DNA to another place. Such element's insertion lead to mutation which is referred as transposon mutagenesis.

» Insertion sequences: The simplest type of transposable sequence (about 700 – 1000 bp) which has the transposase gene alone, which causes the mutation.

» Mu phage: This double strand DNA containing temperate bacteriophage can able to act as transposable elements, causes mutation by transposition.

Site directed Mutation

So for we have discussed about the mutations which occur at random, that means, the gene which is mutated and type of mutation cannot be predicated, referred a random mutation.

The present modern molecular techniques allow to mutate an organism at specifi portion of gene by means of special vectors (plasmids) with controlled system. Thi type of mutation is referred as site directed mutation.

Genetic Recombination

Genetic recombination involves the physical exchange of genetic material betweer organisms. It involves the genetic exchange between homologous DNA sequence form two different organisms. In classical genetics, it is referred as crossing over In prokaryotes, ReaA protein is a DNA binding protein involved in the genetic recombination. The following sequence of steps involved in the genetic recombinatior of prokaryotes.

The cell which gives DNA is donor and which receives the DNA is recipient In prokaryotes, the genetic recombination is observed because of homologous DNA from one cell to another by three processes.

1. Transformation: Donor DNA free in the environment (naked DNA) transferred to recipient cell.
2. Transduction: Transfer of donor DNA to recipient cell by means of virus.
3. Conjugation: Transfer of donor DNA to recipient cell by cell to cell contact

1. Transformation

A naked DNA or cell free DNA incorporated into the recipient cells and brings about agenetic recombination is referred as transformation.

Most of the bacteria both C+ and G- are easily transformable, but only a fragment of DNA can easily be transformed (few number of genes).

Genetic recombination by transformation

The ability of an organism to take up the DNA molecule is referred to as competence, and the organism is said to be competent. The competence of organisms varies among the organisms. Bacillus has 20% competent cells, and E. coli and Streptococcus, 100% of cells become competent.

The gram-negative bacterium, Haemophilus, takes only double-stranded DNA, whereas the Streptococcus and Bacillus can take only single-stranded DNA while the complementary strand is simultaneously degraded.

The naked DNA first bound on the surface of the recipient cell with the help of surface proteins. A single strand of DNA is allowed to enter inside the cell. The entered DNA will be moved to chromosomal DNA, and genetic recombination takes place with the help of RecA protein.

Artificial transformation

The molecular biology techniques now used to transfer any naked DNA (in the form of vectors) to recipient cells are artificially.

1. Calcium chloride mediated transformation: The CaCI2 washed recipient cells mixed with DNA allowed a temperature shock (420C for 40-60 sec.) will allow the naked DNA to get transformed.
2. Gene gun (Particle gun): will allow transformation of plant tissues, yeast, algae etc.
3. Electroporation: Transformation takes place by means of electrical pulses.

2. Transduction

A DNA transferred from one cell to another through the agency of virus is referred as transduction. The transduction of bacteria occurs through bacteriophages.

Some phages are lytic, which kill the host cell during replication, called lytic cycle and lytic phages. Ex. T2 phage; Some phages replicate in the bacteria without killing the host cell, called as lysogenic cycle and temperate phage. The host cell is called as lysogen.

Based on the behavior of virus, transduction occurs in two way. 1. Generalized transduction and 2. Specialized transduction.

Generalized transduction

The generalized transduction occurs due to lytic phage. Normally, the lytic phage sends its DNA after adsorption into the host cell, and DNA will order the chromosomal DNA to synthesize viral DNA and viral proteins. Then packing takes place, and virions lyse the host cell and escape to the environment. Accidentally, sometimes, the viral protein packed with bacterial DNA instead of viral DNA during the lytic cycle is referred to as a transducing particle. This transducing particle is able to infect the host cell but cannot perform the normal lytic cycle. When the transducin particle infects the recipient cell, the DNA will penetrate into the cytoplasm anc with the help of RecA protein, get into chromosomal DNA, which leads to geneti recombination.

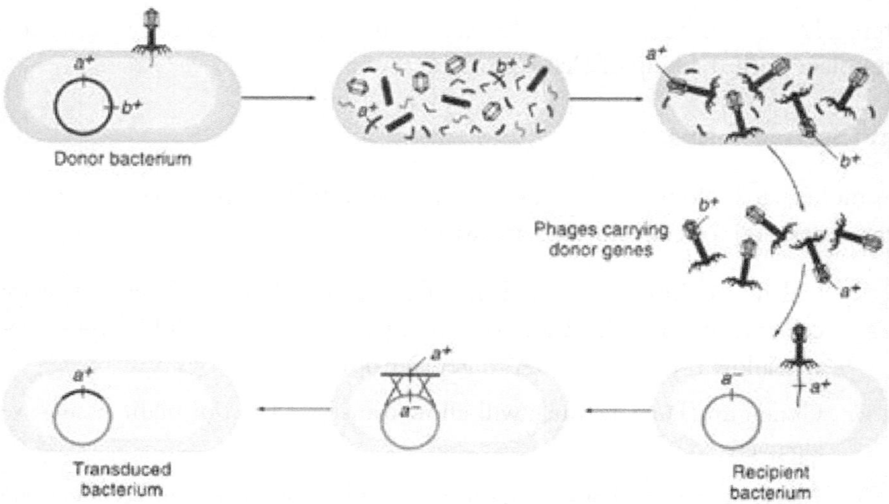

Fig 12.1: Generalized transduction

The frequency of transduction is 1 in 106 to 108 cells. This method of transductio is generalized one. Any portion of donor DNA can become transducing particle anc become transduced transduction takes place; this type of transduction is referred a generalized transduction.

Specialized transduction: On the other hand, some temperate phages can abl to take up a specific portion of donor DNA as their part and transduction take place, which is referred as specialized transduction.

The temperate phage (Lambda phage) send its DNA after adsorption. Th penetrated viral DNA will get integrated into bacterial chromosomal DNA anc persist for ever. This stage of phage is referred as prophage. The cycle is lysogenei cycle. Due to chemical or physical induction, the prophage get activated, replicatec its viral DNA and protein coat followed by assembly and lysis of the cell.

The lambda phage will get integrated into bacterial chromosome at specifi region (near to gal gene – galactose operon). When induce the prophage, sometimes the viral DNA wrongly included some portion of bacterial DNA (mostly gel gene with loss of protion of viral DNA. After assembly, these phages will escape tc ıvironment. These phages are designated as lambda (dgal) phages. These λ(dgal) ıages cannot act as that of normal lambda phages. For infection of recipient cells, ıe λ(dgal) phages need the help of normal lambda phages (referred as helper phages). Vhen these λ(dgal) phages infect the E. coli (gal defective gal), the transduction ke place lead to gain of gal gene. The recipient cells become gal^+.

. Conjugation

acterial conjugation (mating) is a process of genetic recombination that involves ell to cell contact.

;enetic Recombination by Conjugation

he gene for conjugation (ability to mate) is a plasmid-borne character. The gene ıvolved in the conjugation is designated as the F factor (fertility factor), which is ›cated in the plasmid (referred to as the F plasmid). The cell that has the F plasmid is ıe donor (male), designated as F+ cells, and the cell that doesn't have the F plasmid the recipient (female), designated as F- cells. The DNA will be transferred from onor to recipient through physical contact. So another requirement for cell-to-ell contact is a sex pilus through which the donor DNA will be transferred to the ecipient cell.

The sex pili makes a connection between donor and recipient. A nick (split) ı the complementary strand of the F plasmid takes place. The single-stranded lasmid transferred to the F- cell and became double-stranded. The F+ cell also ynthesized its cDNA and became double-stranded. After conjugation, both the ells become F+ cells.

Ifr cells

ometimes, F plasmid transferred from F+ cell to F- cell get integrated into ecipient cell chromosomal DNA. Such conjugants possess the F factor in the hromosomal DNA instead of plasmids. They are referred as hfr cells (High Frequency Recombination). They have very high frequency of conjugations.

)ther genetic recombinations

he process by which a gene moves from one place of chromosome to another in genome is called as transposition. The genes involved are transposon genes or ransposable elements. Insertion sequences are also a type of transposable element of about 700 bp size) which also can cause the genetic recombinations. These

DNA are used to insert a foreigh DNA to an organism and also cause mutagene (discussed earlier).

Genetic Engineering

Development of sophisticated procedures for isolation, manipulation and expressi of genetic materials in organism, a field is called as genetic engineering. Isolatic purification and replication of fragment of DNA through some vectors are referr as gene cloning. The plasmids and bacteriophages are used as vector for gene clonir

Chapter - 13

Industrial Microbiology

Industrial Microbiology is the discipline that uses micro organisms, usually grown on large scale, to produce commercial products or carry out important chemical transformations.

The products derived from such processes are

» Alcohols and alcoholic beverages

» Production of pharmaceutical agents like anitibiotics, vaccines etc.

» Food additives

» Enzymes

» Chemicals

» Single cell proteins

All these industrial microbiological processes are enhancement of metabolic reactions of microorganisms with a goal of over production of these products.

Now the microbial biotechnology joined with the industrial microbiology, by means of genetic manipulations of these microorganisms lead to many fold increase of products and also to produce new products, which are originally not produced by these microorganisms.

Products of microorganisms can be grouped into three, based on their activity.

» As cell: The microbial cell itself as a product of the industrial process. Ex. Baker yeast, Single cell proteins like yeast, Chlorella etc.

» Bioconversion: Microorganisms act as a biocatalyst to convert one substrate to another value added product. Ex. Steroids

- Products from cell: The products will be separated from microorganisms afte the large scale production is over. Ex. Alcohols, antibiotics, enzymes etc.

Fermentation

In industrial term, fermentation refers to any large scale microbial process carried out either aerobically or anaerobically.

Product formation

The product formation during fermentation (or microbial growth) was of two groups

1. Primary metabolites: The product formed during growth of an organism. I means that the product will be formed during the active growth stage (log phase) of the microorganism. Ex. Alcohol by yeast. Since alcohol is produced by the primary metabolic pathway of yeast (EMP pathway), the alcohol wil be accumulated during the active growth stage of yeast.
2. Secondary metabolites: The products formed at the end of active growth stage or at the beginning of the metabolites. Ex. Antibiotics are produced usually at the end of log phase. These products were formed by the secondary metabolic pathways which are not essential for its growth. These kinds of metabolites are produced only by few organisms.

By altering the conditions and strain improvement, it is possible to get over production of secondary products, whereas it is difficult to get over production of primary metabolites.

Type of Fermentation

Based on the stage, the fermentation can be classified in to two.

Submerged (liquid) fermentation: Fermentation is carried out as liquid forms Ex. Alcohol from sugarcane industrial waste – Molasses by yeast

Solid fermentation: Fermentation is carried out as solid condition with about 30-50% moisture level. Ex. Gibberelic acid production from grains by fungus Fusarium *miniliformi.*

Based on the microbial growth (culturing method):

- Batch fermentation: A closed system of microbial fermentation in which the volume is fixed.

» Fed batch fermentation: The substrate for fermentation (raw material) is added in increments as the fermentation progresses.

» Continuous fermentation: An opened system of fermentation in which the input (nutrient solution) is added continuously with a withdrawal of equal portion of converted or fermented solution.

Large-scale fermentations

The vessel in which industrial fermentation process carried out is called as fermentor. The volume of the fermentor varies from 5-10 lit to 500,000 lit. Similarly, the fermentor model will vary for aerobic process and for anaerobic process.

The basic structure of fermentor is as follows:

The components are

» Inlet: to feed the substrates, microorganism etc.

» Outlet: to collect the products, to take samples

» Stirrer: to mix the contents so as to get contact between nutrients and microbes

» Air sparger: to provide oxygen to the microbes by small air bubbles.

» pH meter/controller: to check the pH and if needed to add acid and alkali to maintain optimum pH

» Thermometer: to check the temperature

» Outer jacket: by which cool water will be circulated to cool the fermentor (because during fermentation, lot of heat will be generated)

» Sterilization unit: by which steam will be supplied to the fermentor to sterilize the medium and container

» Scale up: Scale up refers the step by step process in any fermentation to get the maximum product.

Fermentor – Model

Scale up of fermentation

Products from Industrial Microbial Fermentations

The following list show the commercial products obtained through fermentatio
by different microorganisms.

a. Antibiotics

Most of the antibiotics are produced directly though microbial fermentation: Sometimes, the antibiotic produced by an organism will be again modified throug chemical process, referred as semi-synthetic antibiotics. The table shows som antibiotics and the organisms involved in commericial production.

Antibiotics	**Organism**
Bacitracin	Bacillus *lichenniformis*
Cephalosporin	Cephalosporium *sp.*
Streptomycin	Sreptomyces *griseus*
Kanamycin	Sreptomyces *kanamycetus*
Erythromycin	Sreptomyces *erytheus*
Polymyxin	Bacillus *polymyxa*
Penicillin	Penicillium *chrysogenum*

b. Vitamins

Vitamin B12 and riboflavin are commercially produced by microorganisms Propionibacterium and Pseudomonas denitrificans produce vitamin B12 and fung Ashbya gossipii produce riboflavin.

c. Amino acids

L-glutamate, L-aspartate, glycine, cysteine, tryptophan, aspartate, monosodiun glutamate (MSG) are common amino acids commercially produced by microbes Presently, the genetically modified bacteria are used to produce these amino acids

d. Bioconversion

In bioconversion process, the microorganisms act as biocatalyst for conversion of on compound to another product. Ex. Most of steroids are produced by bioconversion process.

. Enzymes

he enzymes are mostly used for food industry and molecular biology. The following ıble shows the enzymes and responsible organisms.

Enzyme	Organism
Amylase	Bacillus licheniformis, Aspergillus
Protease	Bacillus licheniformis
Pectinase	Aspergillus niger
Glucose isomerase	Bacillus coagulans
Rennin	Streptococcus lactis
Lipase	Aspergillus, Candida
DNA polymerase	Thermos aquaticus

Chemicals

'inegar (acetic acid), citric acid, acetone, butanol etc. are some examples of the hemicals produced by microorganisms.

Vinegar – Acetobacter – Gluconobacter; Lactic acid – Lactobacillus; Citric cid – Aspergillus niger, Acetone – Clostridium acetobutylicum.

. Single cell protein

he proteins produced by microorganism as food is referred as single cell proteins (SCP). he yeast, Chlorella, cyanobacteria, Spirulina are some common examples of SCPs.

. Alcohols and alcoholic beverages

he yeast play a major role in the alcohol industry. Since, it is a facultative aerobe, can e grown aerobically and anaerobically. When, yeast grown aerobically, its biomass cell yield) increased with low alcohol, used for production of baker yeast. Whereas, east grown anaerobically, the alcohol production will be maximum.

Brewing refers the manufacture of alcoholic beverages from fermentation of ıalted grains. Wine is obtained form grape; beer from barley; whiskey, rum, otka tc are the commercial preparations of alcohols. The beer is differentiated form other everages by its processing method. After fermentation, the beer is prepared as such ithout distillation, instead, pasteurization and packing gives beer. Whereas other ommercial beverages are prepared after distillation of alcohol.

i. Food

The microorganisms as such can be used as food and by their action also many food
are produced. Following are some of the foods obtained form microbes.

- » Mushroom: Agaricus bisflorus (button mushroom); Pleurotus sajor-ca (oyster mushroom).
- » Bread: by yeast
- » Butter, yogurt, like dairy products: Lactic acid bacteria (Lactobacillu Leuconostoc).

j. Vaccines

The vaccines are the immunized antigens which has whole cell or part of cell or ce product. The attenuated virus is a best example for whole cell virus vaccine; Present the viral DNA also act as vaccine which is referred as DNA vaccines. All these a commercially produced.

Chapter - 14

Immunology - Host Defence Mechanism

The pathogenicity of an organism varies form mild attack to severe fatal attacks. The host has many mechanisms to resist or to avoid the pathogens apart from immune systems. One or few of these will be activated when the pathogen enters into the body. Following are some of the listed natural host defense mechanisms.

1. Natural host resistance: The host has the natural ability to resist the pathogen. For example, the anthrax may cause mild pustules to man whereas cause fatal blood poisoning to cattles.
2. Age, stress and diet: Very young and very old are susceptible whereas the adults are generally resistant against pathogens. Similarly the physical stress induces the susceptibility of the host against pathogens. Malnutrition is also allowed the pathogens to act very rapidly.
3. Anatomical defenses: The body of host (man) has so many defense mechanisms to avoid the pathogens. Lysozyme of tears; rapid passage of air in nose; highly acidic pH of stomach (pH2); rapid change in pH of intestinal track; native microflora of large intestines; skin layers; mucous membrances of lungs; coughing; sneezing are some anatomical defenses of our body to resist the pathogens.
4. Fever and inflammation: Redness, swelling, pain, heat and fever are some non-specific reactions carried out by host against pathogens.
5. Immune system: The specific and non-specific immunity is the well-advanced resistance mechanism, which will be discussed in this chapter.

Immunology

Immunology is the study deals with the mechanism by which an organism able to resist the infection by pathogens.

Immunity is defined as the ability of an organism to resist the pathogens infection. The immunity can be classified in to two groups.

1. Non-Specific immunity – refers the general ability of certain cells of the body to resist the most of the pathogenic organisms like bacteria, fungi, virus etc
2. Specific immunity – refers highly sophisticated mechanism found in the vertebrates for developing resistance to individual pathogens.

These two immune systems were triggered when the foreign molecule (pathogenic microorganisms or their toxins) entered the cell. These foreign molecules macromolecule components of pathogens, such as surface proteins and excreted proteins (toxins), are collectively referred to as immunogens or antigens.

In response to an antigen, the host immune system will produce soluble proteins to detoxify the antigen's effect, which is referred to as an antibody.

Once an antibody is produced by an immune system for a particular antigen the second time an infection by the same microorganism or the same antigen occurs it will be resisted by rapid antibody production. This capacity of the immune system for responding to the challenge of additional exposure to a pathogen or antigen is referred to as immune memory.

Blood and Lymph

Blood and lymph are the two major body fluids responsible for all these immune system.

Blood consists of both cellular and acellular components.

Cellular components

- Erythrocytes (red blood cells) – Non nucleated cells; major function is the O_2 carrier to cells
- Leucocytes (white blood cells) – involves in the immunity; two kinds are present
- Phagocytes / monocytes: They are the phagocytic cells (eating cells) which will engulf the pathogens
- Lymphocytes: responsible for antibody production against antigens

Acellular components

- » Plasma – When the cellular components removed from blood, is referred as plasma. It has an important protein called fibrinogen, which is responsible for blood clotting.
- » In plasma, if the blood clotting protein is removed, that fluid is referred as Serum.

Lymph: is a fluid similar to blood but lack of red blood cells and travels through separate circulatory system (the lymphatic system). This system has capillary like arrangement called lymph node to filter the pathogenic cells.

Antigens or Immunogens

They are the substances that, when admitted to a host (vertebrates), are able to induce the immune system of the body, especially the specific immune system, and are referred to as antigens or immunogens. These include a variety of macromolecules of pathogens, namely many polysaccharides, surface proteins, nucleic acids, lipoproteins, etc.

- » Some times, the organism especially bacteria secrete protein extracellularly, which produce immediate damage to the host cell are referred as exotoxins. Such exotoxins released at small intestine of host are referred as enterotoxins.
- » Sometimes, the bacterial cell wall lipopolysaccharides act as toxin when solubilized, which is referred as endotoxins – All these are act as antigens.

Non-specific Immunity

Refers to the general ability of certain cells of the body to resist the most of the pathogenic organisms like bacteria, fungi, viruses, etc. The non-specific immunity is a common resistance for all pathogens, whereas the specific immunity is specific to every pathogen. The common non-specific immunity is phagocytosis. (engulfing the pathogens).

Phagocytes and Phagocytosis

Some leukocytes (white blood cells) found in the cell are phagocytes. They are motile, amoeboid actions that contain lysosomes. By chemical attraction, the phagocytes move towards pathogens and engulf them. By the action of lytic enzymes, they kill the pathogens. This phenomenon is referred to as phagocytosis.

There are two groups of phagocytes in blood.

» Neutrophiles – short lived (2-3 days), small sized, large numbered phagocyte
They can move very fastly (40 mm/min.).

» Monocytes & macrophages – Large sized phagocytes which live for long tim
Macrophages are fixed in tissues and monocytes are circulating in blood.

» During phagocytosis

» Phagocytes changed their metabolism from aerobic to anaerobic form

» Formation of lactic acid drastically reduced the pH to 2

» Hydrolytic lysozymes excreted

Due to all these causes, the pathogenic organism will be killed. Sometime bacterial pathogens show resistance against phagocytes because of the presence o capsules.

Specific Immunity

Refers to the ability of the immune system to interact with a specific antigen, whic is referred to as specific immunity. The white blood cells (lymphocytes, a type o leucocytes) are responsible for such specific immunity. These lymphocytes produc soluble proteins against antigens, which can interact with the antigens and neutraliz the effect, referred to as antibodies.

The lymphocytes are of two types based on their action and origin. They ar B lymphocytes (shortly as B cells) and T lymphocytes (T cells). B lymphocyte developed from bone marrow cells (designated as B cells), and T lymphocyte developed from the thymus (designated as T cells).

B Lymphocytes

» They are really responsible to produce antibody against antigens and als for immune memory

» The antibody produced by B cells are referred as immunoglobulins (lg).

T Lymphocytes

» They have Antigen Specific T Cell Receptor (ATC), similar to antibody o B cells, which can interact with antigens.

hey have multiple functions such as:

- They recognize antigen and facilitate B cell to act on the antigen
- They recognize antigen and release ATC which can suppress the antigens.

ntibodies

nmunoglobulins, ATCs and cytokines are the proteins produced by B cells and T ells respectively against antigens, which can be referred as antibodies.

. Immunoglobulins (IG)

- They are the soluble protein produced by T cells against antigens
- They are found in the serum, and the serum rich of antibodies is referred as antiserum.
- The immunoglobulins are the circulating antibodies
- Molecular weight ranges from 150,000 to 970,000 daltons
- Five classes of imunoglobulins are recongnized (Ig G, Ig M, Ig D, Ig E and Ig A)
- All the immunoglobulins (Ig G, M, D, E) present both in blood and serum whereas Ig A is present in saliva.

When the antigen is introduced to the body, the immunoglobulin production owly increases to a level and again slowly declines in the body. This response is eferred to as a primary antibody response. When the second time the same antigen as introduced, the immunoglobulin production suddenly increased (10–100 times iore than the first time). This large raise of antibody production due to a second itroduction is referred to as a secondary antibody response.

A typical immune system response is the production of a broad spectrum of nmunoglobulin molecules against antigens. Normally a single clone of B cell will roduce a single type of immunoglobulin. So, when an antigen is introduced, a eterogeneous population of immunoglobulins will be accumulated in the serum, hich is referred to as polyclonal antiserum.

However, it is possible to produce a single antibody population (identical in tructure and action) against antigens. This antibody is referred to as a monoclonal ntibody.

- The monoclonal antibodies are used for research and clinical purpose

- Used for therapeutic applications
- Highly useful to cure cancers
- Identification of microorganisms is also possible by monoclonal antibodi

B. Antigen specific T cell receptors (ATCs)

They are the surface proteins produced by T cells. They are present on the surface the T cells, whereas the immunoglobulins are present in serum. These proteins ha short-chained peptides, which can bind with antigens. These proteins are referr to as ATC. The T cells can interact with antigens through ATC. After a compl is formed, these T cells with antigens can move to B cells, phagocytes, or memo cells to deactivate the antigens.

C. Cytokines

A soluble immune response modulator is produced by leucocytes. They are t regulatory soluble proteins produced by lymphocytes during immune response. Th are responsible for signal transduction and send information inside the cell for t rate of metabolism, cell division, protein synthesis, etc.

Various leucocytes against antigens produce these three proteins as speci immune response.

Antigen – Antibody Reaction

The study of antigen – antibody relation in vitro is called as serology. There are many reactions will occur when antigen is reacted with antibody. The comm reactions are as follows:

- Neutralization – The toxin portion of antigen may be blocked by antibo so that it cannot act to cause the disease, referred as neutralization. T antibodies are here referred as antitoxins.
- Precipitation – Generally antibodies have two binding sites. If two differe antigens combined with antibody leads to precipitation of the compl which reduce the toxicity of the antigens.
- Agglutination – If antigen is not soluble, the antibody forms a clumping (ma like insoluble aggregates) when react with that antigen. The agglutinat clumps can be seen under microscopes.

Immunization

So the discussion is mainly about the immunity naturally developed against the pathogen or antigen. This immunity is inherent character, meaning the individual acquired the infection, and that is controlled by the natural immune system. This immune system is referred to as natural active immunity.

On the other hand, it is possible to induce the immune system of an individual by introducing antigens to the body. This process is known as vaccination or immunization. It is an artificial active immunity.

The introducing material, antigen to the body, is known as a vaccine. The material used to induce the artificial immunity of the body, antigen C—a mixture of antigens—is referred to as vaccines.

Following are some antigens used as vaccines:

- Toxoids – Many pathogens produce exotoxins and endotoxins. These toxins are modified and used to induce the immune system, as vaccines are referred as toxoids.
- Ex. Diphtheria, Tetanus toxoids used as vaccines.
- Attenuated cells – It is possible to isolate a mutant strain of pathogen which is avirulent but still has antigen are referred as attenuated strains. These cells when introduced, can induce the immunity without causing disease.
- Ex. Yellow fever, measles, mumps, polio causing viruses can be attenuated and used as vaccines.
- Inactivated cell – Ex. Rabies virus inactivated cells used as vaccines.
- Killed cells – Ex. Typhoids, Plague causing organisms – killed and used as vaccines.
- Polysaccharide of cell – Ex. Pneumonia causing bacterial polysaccharide used as vaccine.
- DNA – Ex. Hepatitis A and B viruses DNA are used after genetic recombination as vaccines.

DNA vaccines

The specific antigen producing gene can be introduced to host cell. The gene produces antigentic protein and stimulates the immune system. This type of vaccines is DNA vaccines.

Artificial Passive Immunity

The above discussed immunity development artificially induces the native immune system. On the other hand, introduction of antibody directly to the body is also possible. The antibodies may be produced using some other mammals like cattle horse etc. by introducing antigen to their body. Then, the antiserum (serum containing antibodies) obtained from cattle may be introduced to the individual. Such developed immunity is referred as passive immunity.

In passive immunity

- The individual will not expose to antigens
- The individual's immune system will not be induced
- No immune memory will be there
- Immunity development is rapid.

Note: (All these characters are just opposite to artificial active immunity).

Chapter - 15

Biotechnology and Genetic Engineering

The development of sophisticated procedures for the isolation, manipulation, and expression of genetic materials is a field called genetic engineering. Genetic engineering has applications in both basic and applied research. In basic research, genetic engineering techniques are used to study the mechanisms of gene replication and expression in prokaryotes, eukaryotes, and viruses. For applied research, genetic engineering is used for the development of microbial cultures capable of producing valuable products such as human insulins, growth hormones, interferons, vaccines, enzymes, etc. The commercial application of genetic sense, biotechnology refers to the use of living organisms to carry out defined chemical processes for industrial application.

The process of isolation, purification, and replication of specific DNA fragments is called molecular cloning or gene cloning. Molecular cloning is the base of most of the genetic engineering procedures. The purpose of molecular cloning is to isolate large quantities of specific genes in pure form. Thus, molecular cloning is very important in genetic engineering and can be divided into several steps depending on the purpose. The steps are as follows:

1. Source DNA / gene: The gene of interest or total genomic DNA or synthesized from RNA template (by the enzyme Reverse transcriptase) or from DNA template (by the enzyme DNA polymerase) using Polymerase Chain Reaction (Known as PCR).
2. Cloning vectors: Bacterial extra chromosomal DNA, plasmids and bacteriophages are used as vectors or vehicles to replicate and carry the source DNA or gene. The DNA will be ligated with the vector and can be replicated and expressed.

3. Host organisms: Thus cloned vectors should be introduced to universal hos organisms (usually bacterium, E. coli) for replication in several copies pe cell and to maintain the cloned vectors.
4. Identification or screening of host clone for desired host bacterial cell wit cloned vector.
5. Production of large number of desired host containing vector for isolatio and study the cloned DNA.

Gene Cloning Vectors

1. Plasmids and Bacteriophage Lambda

Plasmids are the extrachromosomal DNA present in most of the bacteria, replicatin independently of the host chromosome. In addition to carrying genes required fo their own replication, most plasmids are natural vectors (gene carriers) because the often carry other genes that confer important properties on their host. Plasmid have very useful properties as cloning vectors. Due to their small size, easy isolatio and manipulation, circular DNA, independent origin of replication, and multipl copy number in host cells lead plasmids as ideal vectors. The plasmids pUC18 pUC19, and pBR322 are some of the common plasmid vectors used for commo molecular cloning.

Similarly, modified lambda bacteriophages are highly suitable for gene cloning Wild Lambda phage harbors too many restriction enzyme sites, which is not suitabl as a cloning vector. The unwanted restriction enzyme sites removed by point mutatio called Charon phage can be used as a vehicle for inserting any foreign DNA. Charo 4A and Charon 16 are some vector lambda phages used in gene cloning.

2. Cosmids

Cosmids are plasmid vectors containing foreign DNA with cos site from Lambd genome. These cos sites are required for packaging DNA into Lambda virions. Th main advantages of use of cosmids is to clone large fragment of DNA.

3. Phagemids

Phagemids are the vectors synthetically designed using basic genes from both plasmi and bacteriophages. Most of the modern vectors commercially used in gene cloning are phagemids.

xpression vectors

ı earlier times, vectors were designed mainly to clone the foreign DNA and to ıaintain these cloned vectors in universal host cells (commonly called cloning vectors). ater, the vectors are designed to express the foreign DNA in host cells in terms f the synthesis of proteins coded by the DNA. These vectors are called expression ectors. These vectors allow the foreign genes to express in host cells, overproduce the rotein, and excrete them from the cell. In these vectors, the complete structural genes f the protein will be cloned in the vector with the host-specific promoter sequence.

huttle vectors

ı advanced genetic engineering research, it will be useful to be able to move the enetic materials between completely unrelated organisms. For this purpose, shuttle ectors are used. It is one that can replicate in two different organisms. Shuttle ectors are developed that replicate in both E. coli and Bacillus and yeast, E. coli, nd mammalian cell culture like a pair of organisms.

Iost for cloning vectors

he ideal characteristics of a host plant for cloned genes are rapid growth, capability f growth in inexpensive culture medium, not being harmful or pathogenic, the bility to take up the foreign DNA (competence), and stability in culture. The host ıust have appropriate enzymes to allow replication of the vectors. The most useful osts for cloning are microorganisms that grow well and for which we have both ufficient genetic information and tools for genetic manipulation. These include the acteria E. coli, Bacillus subtilis, and Saccharomyces cerevisiae.

election of Clones

\ crucial step in gene cloning is finding the host that harbor the "cloned vector" referring the plasmid with foreign DNA). For this, two kinds of selection procedures ıave been employed in genetic engineering.

ı. **Selection:** This can allow only hosts containing plasmid vectors by selecting for a vector marker. The most common selection marker is antibiotic-resistant genes incorporated in the vector, which will allow the host to grow in the medium containing antibiotics. This will allow us to differentiate and select the host with the vector and without the vector. But, it will not differentiate the clones containing foreign inserted vectors and clones containing vectors without insert DNA.

b. **Screening:** This procedure allows one to screen several clones to pick the corre clone that harbors the foreign DNA/gene-inserted vector. Inactivation of t β-galactosidase gene (lacZ gene) is the most commonly used screening procedu widely adopted to identify the cloned vector. If the lacZ gene-containing vecto are introduced to host cells and the host cells are allowed to grow in mediu containing the substance 5-bromo-4-chloro-3-indolyl-β-D-galactopyranosi (called x-gal), β-galactosidase produced from the gene lacZ hydrolyzes the x-g and produces the insoluble blue dye, resulting in blue-colored colonies in t plate. If the foreign DNA is inserted in this region (lacZ), then it could not able to utilize the x-gal, and the resulting white colonies will appear. By this w (inactivation of the lacZ gene), the screening of clones can be done to identi the foreign DNA/gene-inserted vectors.

Polymerase Chain Reaction

Polymerase chain reaction (in short, PCR) is used for in vitro rapid amplificati of DNA. PCR can multiply DNA molecules by up to a billionfold in the test tub producing large amounts of specific genes for cloning, sequencing, or mutagenes purposes. DNA polymerase enzyme is used in PCR reactions, which copy the DN

The PCR technique requires a short oligonucleotide, known as a prime complementary to the sequence in the gene of our interest that will anneal in t source DNA (known as template DNA), and with the help of the DNA-depende DNA polymerase enzyme, the complementary DNA strand will be synthesiz The steps in PCR amplification of DNA are as follows:

a. Two primers flanking the target DNA are added in great excess quantity to he denatured target DNA (single stranded DNA).

b. As the mixture cools, the excess of primers related to the target DNA ensur that most target strands anneal to a primer not to each other.

c. DNA polymerase then extends the primers using target strands as template.

d. After appropriate cool, the mixture again heal for denaturation and repeat th above again and again for several cycles.

Thus, each PCR cycle involves "Denaturation", "Annealing" and "Prime extension" resulted several fold synthesis of DNA under in vitro condition.

Application of genetic engineering

Genetic engineering has many commercial and practical applications. The mai applications of biotechnology in the area of interest of commercial development are:

i. Microbial fermentation: Manipulation of microorganisms to overproduce the metabolites suchas aminoacids, antibiotics, etc.

ii. Virus vaccines: As many of the viral vaccines used viral coat proteins as vaccinating materials, genetic engineering of viral coat protein genes cloned in bacteria, nonpathogenic virus or in plants make the viral vaccines development in safe and convenient.

iii. Mammalian proteins: A number of mammalian proteins are greater medicinal and commercial interest are presently allowed to produce by prokaryotic systems in cheaper ways.

iv. Transgenic crops: By introducing cloned DNA directly to the plant cells grown tissue culturally and make expressions known as transgenic crops, are of very economically viable technology. The most popular transgenic crops, known as Bt-crops, which harbor the gene cryIA coding crystal proteins, a toxin against lepidoptran insects, obtained from bacterium, Bacillus thuringiensis. These Bt-transgenic crops are performing better in terms of insect resistance and reducing pesticide applications.

v. Environmental Biotechnology: Because of the enormous metabolic diversity of bacteria, a large pool exists in bacteria from natural habitats. Many bacteria able to produce metabolites / proteins against environmental pollutants. Such genetic sources are presently used with the help of gene cloning to detoxify the pollutants, wastes, waste treatments, degradation of recalcitrant chemicals suchas pesticides, toluenes, anilines, hydrocarbons etc.

vi. Gene therapy: Controlling the genetic disorders and diseases by using gene cloning / genetic engineering is present advanced medical research known as "Gene therapy".

Chapter - 16

Environmental Microbiology

Microorganisms live as single cells in nature in association with other cells in an assemblage that we call a "population.". Such populations are composed of groups of related cells generally derived by successive cell divisions from a single parent cell. The location in an environment where a population generally lives in association with other populations of cells is called "microbial communities." Environmental microbiology deals with the study of these microbial communities in different habitats, naturally occurring on earth.

Natural habitats of microorganisms are exceedingly diverse. Any habitat that is suitable for the growth of microorganisms can also support the growth of other microorganisms. But in addition, there are many habitats where, in extreme environments, due to chemical and physical factors, higher organisms cannot grow, but microorganisms can grow and flourish. In this chapter, different habitats of microorganisms are discussed briefly.

Because microorganisms are so small, their habitats are also small. The term microenvironment is often used in ecology to describe where a microorganism actually lives and metabolizes within its habitat.

Terrestrial Environments

Terrestrial environments always refer to the soil and plant habitats. In the soil on or near the plants, many key processes occur that influence the functioning of the ecosystem. The plant roots penetrate into the soil, and their excretions promote the development of rhizosphere microflora.

a. **Soil**: The vast, extensive microbial growth takes place on the surface of the soil particles, usually within the rhizosphere. Major factors affecting microbial activity in soil are the availability of water, organic carbon status, nutrient status, oxygen, pH, and so on. The community structure of microorganisms in soil decides many important elements transformation, which is the key factor for soil fertility in agriculture. They are also a source of soil enzymes, which govern the soil biochemical properties.

). **Deep subsurface microbiology**: The deep soil subsurface terrestrial environment, which can extend for several hundred meters below the soil surface, harbors a variety of microorganisms, primarily bacteria. The bacteria include anaerobes such as sulfate-reducing bacteria, methanogens, and homoacetogens, which are predominant. Compared to surface microorganisms, deep subsurface organisms are low in their metabolic activities. Studying these organisms will be more useful for in situ bioremediation of toxic chemicals. Bioremediation refers to the use of microorganisms to remove or detoxify toxic or unwanted chemicals in an environment.

Aquatic Environments

Typical aquatic environments are the oceans, lakes, ponds, rivers, salt marshes, springs, etc. Aquatic environments differ considerably in chemical and physical properties, and in turn, the microbial community also differs. The predominant phototrophic organisms in most aquatic environments are microorganisms; in oxic areas, cyanobacteria and algae prevail, and in anoxic areas, anoxygenic phototrophic bacteria are predominant. Algae floating or suspended freely in the water are called phytoplankton; those attached to the bottom or sides are called benthic algae. Because these organisms use energy from light in the initial production of organic matter, they are called primary producers.

i. Marine environment: In the oceans, primary productivity is very low due to a limit of nutrients like nitrogen and iron, which limit the phytoplankton population. A consequence of relatively low primary productivity in the oceans is low heterotrophic microbial activity. This in turn ensures that most ocean waters are oxic, even in greater depth.

ii. Deep sea: Deep sea refers to the zone where visible light is absent (below 300 m depth) in the open ocean water. Organisms that inhabit the deep sea are faced with three major environmental factors: low temperature, high pressure, and low nutrient levels. Most of the microbes of the deep sea are psychrophilic, and some are extreme psychrophilic. These organisms must also withstand the enormous hydrostatic pressure associated with great depth.

iii. Lakes and rivers: Significant photosynthetic production of oxygen occurs only in the surface layer of lakes and ponds where light is available. Organic matter is not consumed in this layer; it sinks to the depths and is decomposed by facultative microorganisms using oxygen dissolved in the water. In lakes, once oxygen is consumed, the deep layer becomes anoxic; strictly anaerobic microorganisms alone grow. In the bottom layer of lakes and ponds, where the oxygen and light are almost absent, a highly diversified anaerobic heterotrophic organisms, especially bacterial communities, are predominant. In freshwater ecosystems, the biological oxygen demand (BOD) is the important key factor that decides the microbial community structure.

Extreme Environments and Extremophiles

The world that we are familiar with is full of oxygen, never too cold nor too hot (most of the visible organisms live in a relatively limited range of temperatures from 5^{0}C to 40^{0}C), and we are protected by our atmosphere from most damaging radiation. This familiar world defines what we think of as normal, and the organisms that live at the edges of this world or beyond it are the 'extremophiles.' Interestingly, most extremophiles are microbes, and they may live in places that we might think should be uninhabitable. Such places include the very deep oceans, which are very cold, very dark with no light to support photosynthesis, and under incredible pressure. Or perhaps they can be found in bubbling cauldrons of acidic waters of Yellowstone National Park. Microbial life began probably about 3 billion years ago, and the first evolutionary steps began in a world without oxygen, with a different atmosphere unable to provide protection from ultraviolet radiation, and with water bodies that were hot and acidic. It was a physiologically challenging world, but one in which the microbes survived and succeeded, and through their success they began to change the world—adding oxygen to the atmosphere and changing the microenvironments into places that are less stressed.

The extremes still exist out there, often still dominated by the microbial communities. The study of these environments and the organisms that live there has considerable value for us. The anoxic world (including the sulfur-dominated thiobios) is arguably the most extensive habitat in the world; at best, only the very top layers of sediments and soils are rich in oxygen; the remainder is anoxic. From here, whether rice paddies or mangrove swamps, comes methane, one of the most potent greenhouse gases. As we exhaust those parts of familiar biology that are easily knowable, we move increasingly into the less familiar worlds.

An extremophile is an organism that thrives under "extreme" conditions. The term frequently refers to prokaryotes and is sometimes used interchangeably with rchaea. In this module, however, you will find that extremophiles come in all apes and sizes, and that our understanding of the phylogenetic diversity of extreme abitats increases daily. The term extremophile is relatively anthropocentric. We dge habitats based on what would be considered "extreme" for human existence. Iany organisms, for example, consider oxygen to be poisonous. While oxygen is necessity for life as we know it, some organisms flourish in anoxic environments. /e call them extremophiles.

Most terms used to describe extremophiles are generally straightforward. They e a combination of the suffix – phile, meaning "lover of", and a prefix specific to their ıvironment. For example, acidophiles are organisms that love (phile) acid (acido).

ommon extremophiles are,

1. Acidophile: An organism that grows best at acidic (low) pH values.
2. Alkaliphile: An organism that grows best at high pH values.
3. Anaerobe: An organism that can grow in the absence of oxygen
4. Facultative Anaerobe: An organism that grows in the presence or in the absence of oxygen.
5. Obligate Anaerobe: An organism that cannot grow in the presence of oxygen; the presenc of oxygen either inhibits growth or kills the organism.
6. Endolith: An organism that lives inside rock or in the pores between mineral grains.
7. Halophile: An organism requiring high concentrations of salt for growth.
8. Methanogen: An organism that produces methane from the reaction of hydrogen and carbon dioxide, member of the Archaea.
9. Oligotroph: An organism with optimal growth in nutrient limited conditions.
10. Piezophile (Barophile): An organism that lives optimally at high hydrostatic pressure.
11. Psychrophile: An organism with optimal growth at temperature 15^0C or lower.
12. Thermophile: An organism with optimal growth at temperature 40^0C or higher.
13. Hyperthermophile: An organism with optimal growth at temperature 80^0C or higher.
14. Toxitolerant: An organism able to withstand high levels of damaging elements (e.g., pools of benzene, nuclear waste).

15. Xerophile: An organism capable of growth at very low water activity.

The extreme habitats where these extremophiles thrive are:

» Alkaline: broadly conceived as natural habitats above pH 9 wheth persistently, or with regular frequency or for protracted periods of time.
» Acidic: broadly conceived as natural habitats below pH 5 whether persisten or with regular frequency or for protracted periods of time.
» Extremely Cold: broadly conceived habitats periodically or consisten below 5^0C either persistently, or with regular frequency or for protract periods of time. Includes montane sites, polar sites, and deep ocean habita
» Extremely Hot: broadly conceived habitats periodically or consistently excess of 40^0C either persistently, or with regular frequency or for protract periods of time. Includes sites with geological thermal influences such Yellowstone and comparable Iocations worldwide or deep-sea vents.
» Hypersaline: (high salt) environments with salt concentrations greater th that of seawater, that is, >3.5%. Includes salt lakes.
» Under Pressure: broadly conceived as habitats under extreme hydrosta pressure – i.e. aquatic habitats deeper than 2000 meters and enclosed habita under pressure. Includes habitats in oceans and deep lakes.
» Radiation: broadly conceived as habitats exposed to abnormally high radiatic or of radiation outside the normal range of light. Includes habitats expose to high UV and IR radiation.
» Without Water: broadly conceived as habitats without free water wheth persistently, or with regular frequency or for protracted periods of tim Includes hot and cold desert environments, and some endolithic habitat
» Without Oxygen: broadly conceived as habitats without free oxygen whether persistently, or with regular frequency, or for protracted periods time. Includes habitats in deeper sediments.
» Altered by Humans: heavy metals, organic compounds; anthropogenical impacted habitats. Includes mine tailings, oil impacted habitats.
» Astrobiology: Addresses life beyond the known biosphere – inclusive of li on other heavenly bodies, in space etc. Includes terraforming.

Exercise - 1

Introduction to Microbiology Laboratory and its Equipments

Following are the Instruments

It is used for sterilization of glassware, such as test tubes, pipettes, and petri dishes. Such dry sterilization is done only for glassware. Liquid substances, such as prepared media and saline solutions, cannot be sterilized in an oven, as they lose water due to evaporation. The glassware is sterilized at 180°C for 3 hours. An oven has a thermostat control, using which the required constant temperature can be obtained by trial and error. The thermostat dial reading is approximate, and the exact temperature is read by introducing a thermometer into the oven or on a built-in L-shaped thermometer. In a modern oven there is a digital temperature display and automatic temperature controller to set the desired temperature easily. Time is set by a digital timer. After loading the glassware, the door is closed and the oven is switched on.

The required temperature is set. After the oven attains the set temperature, the required time of sterilization is set on the timer. The oven switches off automatically after the set time. The oven is opened only after its temperature comes down near to room temperature. Otherwise, if the door is opened while the inside of the oven is still very hot, cold air may rush in and crack the glassware's.

1. Drying Oven:

For preparation of certain reagents, the glassware, after proper cleaning and rinsing with distilled water, is required to be dried. They are dried inside the drying oven at 100°C till the glassware dries up completely.

2. Autoclave:

An autoclave is the nucleus of a microbiology laboratory. It is used not only to sterilize liquid substances such as prepared media and saline (diluents) solutions but also to sterilize glassware when required.

It has the same working principle as a domestic pressure cooker. The maximum temperature that can be obtained by boiling water in an open container is 100°C (boiling point of water).

This temperature is sufficient to kill only the non-spore formers, but it is difficult to kill the spore-forming bacteria at this temperature, as they escape by forming heat-resistant spores. It takes a very long time to kill the spores at this temperature.

On the other hand, when water is boiled in a closed container, due to increased pressure inside it, the boiling point elevates, and steam temperature much beyond 100°C can be obtained. This high temperature is required to kill all the bacteria including the heat-resistant spore-formers. Steam temperature increases with an increase in steam pressure.

In operating a standard vertical autoclave, sufficient water is poured into it. If there is too little water, the bottom of the autoclave gets dried during heating, and further heating damages it.

If it has an in-built water heating element, the water level should be maintained above the element. On the other hand, if there is too much water, it takes a long time to reach the required temperature.

The materials to be sterilized are covered with craft paper and arranged on an aluminum or wooden frame kept on the bottom of the autoclave; otherwise, if the materials remain half-submerged or floating, they tumble during boiling, and water may enter. The autoclave is closed perfectly airtight, only keeping the steam release valve open.

Then, it is heated over flame or by the in-built heating element. Air inside the autoclave should be allowed to escape completely through this valve. When water vapor is seen to escape through the valve, it is closed.

Temperature and pressure inside go on increasing. The pressure increase is observed on the pressure dial. Usually sterilization is done at 121°C (a pressure of 15 pounds per square inch, i.e., 15 psi) for 15 minutes. The required time is considered from the point when the required temperature-pressure is attained.

Once the required temperature-pressure is attained, it is maintained by controlling the heating source. After the specified time (15 minutes), heating is discontinued and the steam release valve is slightly opened. If fully opened immediately, due to a sudden fall in pressure, liquids may spill out from the containers.

Gradually, the steam release is opened more and more, so as to allow all steam o escape. The autoclave is opened only after the pressure drops back to normal tmospheric pressure (0 psi). The autoclave should never be opened when there is till pressure inside. The hot sterilized materials are removed by holding them with piece of clean cloth or asbestos-coated hand gloves.

In the case of a steam-jacketed horizontal autoclave, a boiler produces the team. It is released at a designated pressure into the outer chamber (jacket). Air is llowed to escape, and then its steam release valve is closed.

The hot jacket heats the inner chamber, thereby heating the materials to be terilized. This prevents condensation of steam on the materials. Now, steam under ressure is released from the jacket into the inner chamber, and air is allowed to scape from it.

Then, its steam release valve is closed. The steam under pressure in the inner hamber reaches temperatures in excess of 100°C, which can sterilize the materials ept inside it. The autoclave also has an automatic shutting system, i.e., unless the emperature and pressure come down near to room conditions, the door cannot be pened.

Besides the pressure dial, it also has a separate temperature dial to indicate he temperature inside the inner chamber. Moreover, the autoclave maintains he temperature and pressure automatically and switches off after the set time of terilization.

3. Microbiological Incubator:

Profuse growth of microbes is obtained in the laboratory by growing them at uitable temperatures. This is done by inoculating the desired microbe into a suitable ulture medium and then incubating it at the temperature optimum for its growth.

Incubation is done in an incubator, which maintains a constant temperature pecifically suitable for the growth of a specific microbe. As most of the microbes pathogenic to man grow profusely at the body temperature of a normal human being (i.e., 37°C), the usual temperature of incubation is 37°C.

The incubator has a thermostat, which maintains a constant temperature, set according to requirement. The temperature reading on the thermostat is approximate. Accurate temperature can be seen on the thermometer fixed on the incubator. The exact temperature, as per requirement, is set by rotating the thermostat knob by trial and error and noting the temperature on the thermometer.

Most of the modern incubators are programmable, which do not need tri and error temperature setting. Here, the operator sets the desired temperature an the required period of time.

The incubator automatically maintains it accordingly. Moisture is supplie by placing a beaker of water in the incubator during the growth period. A moi environment retards the dehydration of the media and thereby avoids spuriou experimental results.

4. BOD Incubator (Low Temperature Incubator):

Some microbes are to be grown at lower temperatures for specific purposes. Th BOD low-temperature incubator, which can maintain temperatures from 50°C t as low as 2-3°C, is used for incubation in such cases.

The constant desired temperature is set by rotating the knob of the thermosta Rotation of the thermostat knob moves a needle on a dial showing approximat temperature. The exact required temperature is obtained by rotating the knob finely b trial and error and noting the temperature on the thermometer fixed on the incubato

Most of the modern BOD incubators are programmable, which do not nee trial and error temperature setting. Here, the operator sets the desired temperatur and the required period of time. The incubator automatically maintains it accordingl

5. Fridge (Refrigerator):

It serves as a repository for thermo labile chemicals, solutions, antibiotics, serums an biochemical reagents at cooler temperatures and even at sub-zero temperatures (a less than 0°C). Stock cultures of bacteria are also stored in it between sub-culturin periods. It is also used for the storage of sterilized media, so as to prevent thei dehydration.

6. Deep-fridge:

It is used to store chemicals and preserve samples at very low sub-zero temperatures

7. Electronic Top-pan Balance:

It is used for weighing large quantities of media and other chemicals, where precis weighing is not of much importance.

8. Electronic Analytical Balance:

It is used to weigh small quantities of chemicals and samples precisely and quickly

Double-pan Analytical Balance:

is used to weigh chemicals and samples precisely. Weighing takes more time, for hich it is used in emergency only.

). Distilled Water Plant:

/ater is used in the preparation of media and reagents. If the media are prepared ing tap water, the chemical impurities present in it may interfere with the growth the microorganisms in the media. Moreover, the higher the bacteria content of e media, the longer is the time required for their sterilization, and the greater is e chance of survival of some bacteria.

Distilled water, though not bacteria-free, contains fewer bacteria. That is why is preferred in the preparation of microbiological media. It is also used in the reparation of reagents because the chemical impurities present in tap water may terfere with the proper functioning of the reagent chemicals.

The manufacture of distilled water by a Liebig condenser is a time-consuming rocess; in most laboratories, it is prepared by 'distilled water plants.' Usually a istilled water plant is made of steel or brass. It is also called distilled water still.

It has one inlet to be connected to the water tap and two outlets, one for distilled ater to drop into a container and the other for the flow out of hot cooling water to the sink. The still is installed on the wall. It is heated by in-built electric heating ements (immersion heater).

The still works efficiently when the water inflow is so adjusted that the mperature of the cooling water flowing out from the still into the sink is neither o high nor too low; i.e., warm water should flow out. The distilled water may ontain traces of metals corroded from the steel or brass container.

To get metal-free distilled water, glass distillation apparatus is used, and still etter is quartz distillation apparatus. However, for a microbiology laboratory, a eel or glass distillation apparatus is sufficient. For precision analyses, double- or iple-distilled water is used.

1. Ultrapure Water Purification System:

or precision analytical works nowadays, instead of using double- or triple-distilled ater, microfiltered water is used. In the case of distilled water, there is a chance that few volatile substances present in the water get volatilized during the heating of he water and subsequently get condensed into the distilled water collected.

Thus, there may be traces of such substances in the distilled water. To overcor this, ultrapure water is used. Here, water is allowed to pass through very fi microscopic pores, which retain the microscopic suspended particles, including t microbes.

Then, the water passes through two columns of ion exchange resins. The ani exchange resin adsorbs the cations present in the water, while the cation exchan resin adsorbs the anions. The water that comes out is extremely pure.

12. Homogeniser:

For microbiological analysis, liquid samples are directly used, whereas solid sampl have to be mixed thoroughly with a diluent (usually physiological saline) so as get a homogenous suspension of bacteria. This suspension is assumed to conta bacteria homogeneously.

The mixing of solid samples and diluents is done by a homogenizer, in whi a motor rotates an impeller with sharp blades at high speed inside the clos homogenizer cup containing the sample and the diluents. It has a speed regulat for controlling the speed of rotation of the impeller.

In some laboratories, mixing is done manually by sterilized pestle and mort In modern laboratories, a disposable bag is used, inside which the solid sample an liquid diluents are put aseptically and mixed mechanically by peristaltic action of machine on the bag. This machine is called a stomacher.

13. pH Meter:

A pH meter is an instrument for determining the pH of liquid media, liquid sampl and buffers. It has a glass pH electrode. When not in use, it should be kept hal immersed in water contained in a small beaker and preferably be covered by a be jar to avoid dust accumulation in the water and loss of water through evaporatio

Before use, the meter is calibrated using two standard buffers of known pH Usually buffers of pH 4.0, 7.0, and 9.2 are available commercially. The instrument switched on and left for 30 minutes to warm up. The temperature calibration kno is rotated to the temperature of the solutions whose pH is to be measured.

Then, the electrode is dipped into the buffer (pH 7.0). If the reading is n 7.00, the pH calibration knob is rotated till the reading is 7.00. Then, the electroc is dipped in another buffer (pH 4.0 or 9.2).

If the reading is the same as the pH of the buffer used, the instrument is workir properly. Otherwise, the electrode is activated by dipping it in 0.1 N HCl for 24 hours. After calibration, the pH of samples is determined by dipping the electrode into them and noting the reading.

Every time, before dipping into any solution, the electrode should be rinsed with distilled water. The samples should not contain any suspended sticky materials, which may form a coating on the tip of the electrode and reduce its sensitivity.

The old model pH meters have double electrodes (one of them acting as a reference electrode), while new models have a single combined electrode. Moreover, to overcome the problem of temperature correction, now pH meters with automatic temperature correction are available.

Here, another 'temperature electrode' is also put into the solution along with the pH electrode, which measures the temperature of the solution and automatically corrects the influence of temperature variations.

Sophisticated pH meters have a single gel electrode. Such electrodes have very little chance of breakage, as they are almost completely enclosed in a hard plastic casing except at the tip. The tip has both pH and temperature sensors.

Moreover, they are easy to maintain, as they do not require constant dipping in distilled water because the electrode tip is closed with a plastic cap containing a saturated solution of potassium chloride when not in use. However, in preparation of microbiological media, pH is determined by narrow-range pH papers and is adjusted to the required pH by adding acids or alkalis as required.

14. Hot Plate:

Hot plate is used to heat chemicals and reagents. The hot plate is made of an iron plate, which gets heated by an electric heating element from below. The required degree of heating is obtained by a regulator.

15. Shaking Water Bath:

Sometimes, heating at very precise temperatures is required. Such precise temperatures cannot be obtained in an incubator or oven, in which temperature fluctuates, though slightly. However, precise temperatures can be maintained in a water bath, which provides a stable temperature.

A water bath consists of a container containing water, which is heated by electric heating elements. The required water temperature is obtained by increasing or decreasing the rate of heating by rotating the thermostat by trial and error.

In a shaking water bath, the substance is heated at the required temperature and at the same time, it is shaken constantly. Shaking is done by a motor, which rotates and moves the containers to and fro in each rotation. The rate of shaking is again controlled by a regulator. Shaking agitates the substance and enhances th rate of the process.

Most modern water baths are programmable and do not need trial and erro temperature setting. A desired water temperature can be maintained over a desired period of time by programming accordingly. It is used for the cultivation of bacteri in broth medium at a specific temperature.

16. Quebec Colony Counter:

In enumeration of bacteria in samples, it is assumed that a single bacterium give rise to a single visible colony when grown on a plate of solidified nutrient medium. Thus, by counting the number of colonies, the number of bacteria in a sample can be estimated.

Sometimes, colonies are very small and too crowded, making it difficult to count. Counting becomes easy when a mechanical hand counter, called a Quebec colony counter, is used. It divides the plate into several square divisions, and the colonies are magnified 1.5 times by a magnifying glass, which makes counting easy

17. Electronic Colony Counter:

Electronic colony counter is of two types:

1. Hand-held electronic colony counter
2. Table-top electronic colony counter.

The hand-held electronic colony counter is a pen-style colony counter with an inking felt-tip marker. For counting of colonies of bacteria grown in a petri dish, it is kept in an inverted position so that the colonies are visible through the bottom surface of the petri dish.

The colonies are marked by touching the glass surface of the petri dish with the felt-tip of the colony counter. Thus, each colony is marked by a dot made by the ink of the felt-tip on the bottom surface of the petri dish. In a single motion, the electronic colony counter marks, counts, and confirms with a beep sound.

The cumulative count of colonies is displayed on a four-digit LED display. In the case of a tabletop electronic colony counter, the petri dish containing the colonies of bacteria is placed on an illuminated stage, and the count bar is depressed. The recise number of colonies is instantly displayed on a digital readout.

8. Magnetic Stirrer:

n the preparation of solutions, certain chemicals require stirring for a long time to e dissolved in certain solvents. A magnetic stirrer is used to dissolve such substances asily and quickly. A small Teflon-coated magnet, called a 'stirring bar,' is put into container containing the solvent and the solute.

Then, the container is placed on the platform of the magnetic stirrer, below vhich a magnet rotates at high speed by a motor. Attracted by the rotating magnet, he Teflon-coated magnet rotates inside the container and stirs the contents. Now, he solute dissolves quickly.

The Teflon coating prevents the magnet from reacting with the solution, which omes in contact with it. After complete dissolution, the Teflon-coated magnet is emoved from the solution by means of a long retriever, called a 'stirring bar retriever.'.

19. Sonicator:

t is used to rupture cells using high frequency waves.

20. Vortex Mixer:

t is an instrument used for thorough mixing of liquids in test tubes. It has a rotor, vhose speed can be controlled. On the tip of the rotor is a foam-rubber top. When he bottom of a test tube is pressed upon this foam-rubber top, the rotor starts otating, thereby rotating the bottom of the test tube at high speed.

Due to centripetal force, the solution gets mixed thoroughly. It is particularly elpful during serial dilution in the enumeration of bacteria, which needs a omogenous suspension of bacteria cells.

21. Laminar Flow Chamber:

It is a chamber used for aseptic transfer of sterilized materials, as well as for inoculation of microbes. Dust particles floating in the air harbor microbes. These microbe-laden dust particles may enter into the sterilized media and contaminate them when they are opened for short periods of time during inoculation of microbes or transfer from one container to another.

To overcome this, when inoculation is done in open air, the air of the small inoculating area is sterilized by the flame of a Bunsen burner. The heated air becomes light and moves upwards, thereby preventing the dust particles from falling on the media during the short opening process.

To further reduce the chance of contamination by the microbe-laden air, laminar flow chamber is used. It is a glass-fitted cuboidal chamber. An air blowe blows air from the surrounding area and passes it through a HEPA filter (Hig Efficiency Particulate Air filter) so as to make it dust-free (microbe-free).

This microbe-free air passes through the chamber in a laminar manner an comes out from the chamber through the open front door. This laminar flow c microbe-free air from the chamber to outside through the open door prevents th outside air from entering into the chamber.

Thus, the chamber does not get contaminated with the microbes present in th outside air, though the door is kept open during inoculation or transfer of media. UV lamp fitted inside the chamber sterilizes the chamber before operation.

It has a stainless steel platform with provision for gas pipe connection for Bunsen burner. Before use, the platform is cleaned and disinfected with Lysol, th Bunsen burner is connected, and then the glass door is closed.

The UV light is switched on for 10 minutes to sterilize the environment insid the chamber and then switched off. The glass door should never be opened whe the UV light is on, because UV light has a detrimental effect on skin and visior The blower is switched on, and then the glass door is opened.

Now, the Bunsen burner is lighted, and media transfer or inoculation is carrie out in the chamber aseptically. If extremely hazardous microbes are to be handled, laminar flow chamber with gloves projecting into the chamber from the front glas door is used, as inoculation has to be done keeping the front door closed.

22. Electronic Cell Counter:

It is used to directly count the number of bacteria in a given liquid sample. A example of an electronic cell counter is the 'Coulter counter.' In this equipment, suspension of bacterial cells is allowed to pass through a minute orifice, across whicl an electric current flows.

The resistance at the orifice is electronically recorded. When a cell passe through the orifice, being a nonconductor, it increases resistance momentarily. Th number of times resistance increases momentarily is recorded electronically, whicl indicates the number of bacteria present in the liquid sample.

3. Membrane Filtration Apparatus:

ertain substances like urea disintegrate and lose their original properties, if sterilized / heat. Such substances are sterilized by membrane filtration apparatus. In this pparatus, the solution of the substance to be sterilized is filtered through a membrane ter, which does not allow bacteria cells to pass down. Filtration is done under ction pressure to increase the rate of filtration.

4. Microscopes:

ifferent types of microscopes are used for visual observation of morphology, motility, aining and fluorescent reactions of bacteria.

5. Computers:

omputers are generally used for analysis of results. They are also used for identification f bacteria easily within few hours. Otherwise, identification of bacteria is a tedious rocess and takes days together to identify one bacteria species.

The computers used for identification of bacteria are Apple II, IBM PC and RS-80 and their modern variants. Each research personnel of the laboratory should e provided with a computer, along with internet facility.

6. Spectrophotometer:

is an instrument for measuring the differences in color intensities of solutions. A eam of light of a particular wavelength is passed through the test solution and the nount of light absorbed (or transmitted)is measured electronically.

A simple visible spectrophotometer can pass light with wavelengths within sible range, whereas a UV- cum-visible spectrophotometer can pass light with avelengths in ultraviolet as well as in visible range.In microbiology lab, it is used r direct counting of bacteria in suspension as well as for other purposes.

7. Electrical Devices:

fluctuation of electric voltage in the laboratory is one of the most important easons, which reduces the longevity of the equipment and sometimes damages . Therefore, all the voltage-sensitive equipment should be provided with voltage rotection devices like stabilizers, servo stabilizers, or constant voltage transformers CVTs) as per the recommendations of the manufacturers of the equipment.

Computers, balances, and some sophisticated equipment should be connected rough an uninterrupted power supply (UPS), as any breakdown in the electric power ipply during their operation may severely damage some of their sensitive components.

The laboratory should have a high-capacity generator to supply electric current the whole laboratory in case of power failure. This is because a power failure not or brings the activities of the laboratory to a standstill, it also brings about undesirak irreversible changes in the samples stored in the deep freezers and refrigerators.

Exercise - 2

Methods of Sterilization

Sterilization:- It is defined as the process where all the living microorganisms, including bacterial spores are killed. Sterilization can be achieved by physical, chemical and physiochemical means. Chemicals used as sterilizing agents are called chemisterilants.

Physical Methods of Sterilization

Sunlight: The microbicidal activity of sunlight is mainly due to the presence of ultraviolet rays in it. It is responsible for spontaneous sterilization in natural conditions. In tropical countries, the sunlight is more effective in killing germs due to a combination of ultraviolet rays and heat. By killing bacteria suspended in water, sunlight provides a natural method of disinfection of water bodies such as tanks and lakes. Sunlight is not sporicidal; hence it does not sterilize.

Heat: Heat is considered to be the most reliable method of sterilization of articles that can withstand heat. Heat acts by oxidative effects as well as denaturation and coagulation of proteins. Those articles that cannot withstand high temperatures can still be sterilized at lower temperatures by prolonging the duration of exposure.

Red heat: Articles such as bacteriological loops, straight wires, tips of forceps, and searing spatulas are sterilized by holding them in Bunsen flame till they become red hot. This is a simple method for effective sterilization of such articles, but it is limited to those articles that can be heated to redness in flame.

Flaming: This is a method of passing the article over a Bunsen flame but not heating it to redness. Articles such as scalpels, mouths of test tubes, flasks, glass slides, and cover slips are passed through the flame a few times. Even though most vegetative cells are killed, there is no guarantee that spores too would die on such short exposure. This method too is limited to those articles that can be exposed toflame. Cracking of the glassware may occur.

Hot air oven: This method was introduced by Louis Pasteur. Articles to be sterilized are exposed to high temperature (160°C) for a duration of one hour in an electrically heated oven. Since air is a poor conductor of heat, even distribution of heat throughout the chamber is achieved by a fan. The heat is transferred to the article by radiation, conduction, and convection. The oven should be fitted with a thermostat control, a temperature indicator, and meshed shelves and must have adequate insulation.

Articles sterilized: Metallic instruments (like forceps, scalpels, scissors), glassware (such as Petri-dishes, pipettes, flasks, all-glass syringes), swabs, oils, grease, petroleum jelly and some pharmaceutical products.

Autoclave: Sterilization can be effectively achieved at a temperature above 100°C using an autoclave. Water boils at 100°C at atmospheric pressure, but if pressure is raised, the temperature at which the water boils also increases. In an autoclave, the water is boiled in a closed chamber. As the pressure rises, the boiling point of water also raises. At a pressure of 15 lbs inside the autoclave, the temperature is said to be 121°C. Exposure of articles to this temperature for 15 minutes sterilizes them. To destroy the infective agents associated with spongiform encephalopathy (prions), higher temperatures or longer times are used; 135°C or 121°C for at least one hour are recommended.

Filtration

Filtration does not kill microbes; it separates them out. Membrane filters with pores sizes between 0.2-0.45 μm are commonly used to remove particles from solutions that can't be autoclaved. It is used to remove microbes from heat-labile liquids such as serum, antibiotic solutions, sugar solutions, and urea solutions. Various applications of filtration include removing bacteria from ingredients of culture media, preparing suspensions of viruses and phages free of bacteria, measuring sizes of viruses, separating toxins from culture filtrates, counting bacteria, clarifying fluids and purifying hydrated fluid. Filtration is aided by using either positive or negative pressure using vacuum pumps. The older filters made of earthenware or asbestos are called depth filters.

Exercise - 3

Nutritional Media and their Preparations

Nutrient Agar is a general purpose, nutrient medium used for the cultivation of microbes supporting growth of a wide range of non-fastidious organisms. Nutrient agar is popular because it can grow avariety of types of bacteria and fungi, and contains many nutrients needed for the bacterial growth.

Composition of Nutrient Agar

» 0.5% Peptone

It is an enzymatic digest of animal protein. Peptone is the principal source of organic nitrogen for the growing bacteria.

» 0.3% beef extract/yeast extract

It is the water-soluble substances which aid in bacterial growth, such as vitamins, carbohydrates, organic nitrogen compounds and salts.

» 1.5% agar

It is the solidifying agent.

» 0.5% NaCl

The presence of sodium chloride in nutrient agar maintains a salt concentration in the medium that is similar to the cytoplasm of the microorganisms.

» Distilled water

Water is essential for the growth of and reproduction of micro-organisms and also provides the medium through which various nutrients can be transported.

» pH is adjusted to neutral (7.4) at 25 °C.

Preparation of Nutrient Agar

1. Suspend 28 g of nutrient agar powder in 1 litre of distilled water.
2. Heat this mixture while stirring to fully dissolve all components.
3. Autoclave the dissolved mixture at 121 degrees Celsius for 15 minutes.
4. Once the nutrient agar has been autoclaved, allow it to cool but not solidif
5. Pour nutrient agar into each plate and leave plates on the sterile surfac until the agar has solidified.
6. Replace the lid of each Petri dish and store the plates in a refrigerator

Uses of Nutrients Agar

1. It is frequently used for isolation and purification of cultures.
2. It can also be used as a means for producing the bacterial lawns neede for antibiotic sensitivity tests.In actuality, antibiotic sensitivity testing i typically performed on media specially formulated for that purpose.

xercise - 4

Microscope - Parts, Principles of Microscopy, Resolving Power and Numerical Aperture

high-power or compound microscope achieves higher levels of magnification an a stereo or low-power microscope. It is used to view smaller specimens, such cell structures, which cannot be seen at lower levels of magnification. Essentially, compound microscope consists of structural and optical components. However, ithin these two basic systems, there are some essential components that every icroscopist should know and understand. These key microscope parts are illustrated id explained below.

tructural Components

he three basic, structural components of a compound microscope are the head, ase and arm.

- Head/Body houses the optical parts in the upper part of the microscope
- Base of the microscope supports the microscope and houses the illuminator
- Arm connects to the base and supports the microscope head. It is also used to carry the microscope. When carrying a compound microscope always take care to lift it by both the arm and base,simultaneously.

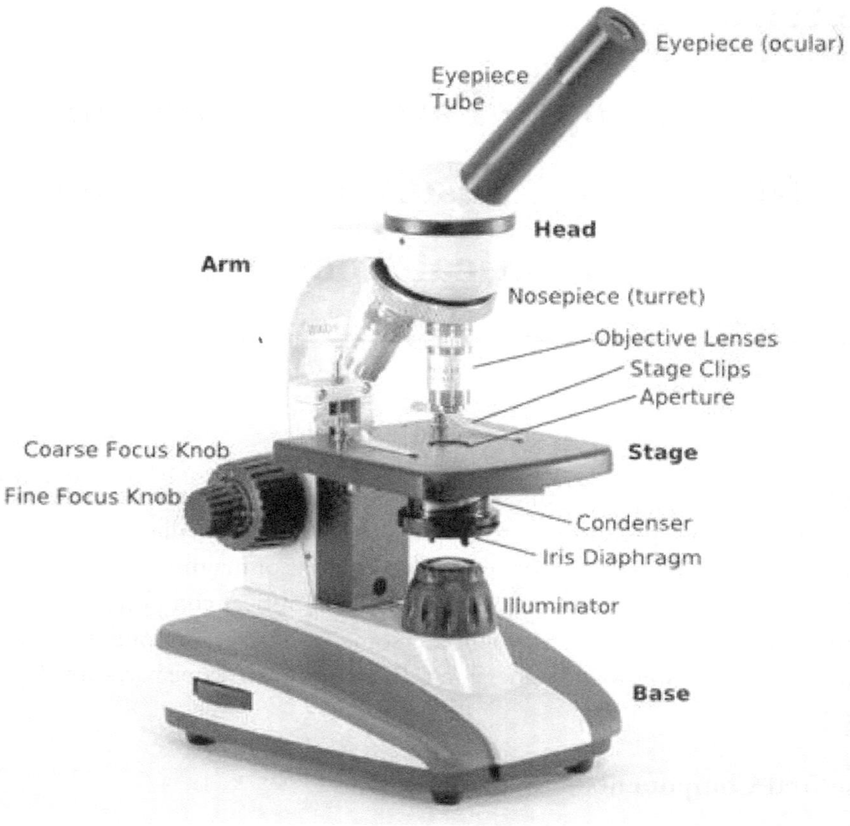

Optical Components

There are two optical systems in a compound microscope: Eyepiece Lenses ar Objective Lenses:

Eyepiece or Ocular is what you look through at the top of the microscope. Typical standard eyepieces have a magnifying power of 10x. Optional eyepieces of varyin powers are available, typically from 5x- 30x.

Eyepiece Tube holds the eyepieces in place above the objective lens. Binocul microscope heads typically incorporate a diopter adjustment ring that allows for th possible inconsistencies of our eyesight in one or both eyes. The monocular (sing eye usage) microscope does not need a diopter. Binocular microscopes also swiv (Interpupillary Adjustment) to allow for different distances between the eyes different individuals.

Objective Lenses are the primary optical lenses on a microscope. They range from 4x-100x and typically, include, three, four or five on lens on most microscopes. Objectives can be forward or rear-facing.

Nosepiece houses the objectives. The objectives are exposed and are mounted on a rotating turret so that different objectives can be conveniently selected. Standard objectives include 4x, 10x, 40x and 100x although different power objectives are available.

Coarse and Fine Focus knobs are used to focus the microscope. Increasingly, they are coaxial knobs - thatis to say they are built on the same axis with the fine focus knob on the outside. Coaxial focus knobs are more convenient since the viewer does not have to grope for a different knob.

Stage is where the specimen to be viewed is placed. A mechanical stage is used when working at higher magnifications where delicate movements of the specimen slide are required.

Stage Clips are used when there is no mechanical stage. The viewer is required to move the slide manually to view different sections of the specimen.

Aperture is the hole in the stage through which the base (transmitted) light reaches the stage.

Illuminator is the light source for a microscope, typically located in the base of the microscope. Most lightmicroscopes use low voltage, halogen bulbs with continuous variable lighting control located within the base.

Condenser is used to collect and focus the light from the illuminator on to the specimen. It is located under the stage often in conjunction with an iris diaphragm.

Iris Diaphragm controls the amount of light reaching the specimen. It is located above the condenser and below the stage. Most high quality microscopes include an Abbe condenser with an iris diaphragm. Combined, they control both the focus and quantity of light applied to the specimen.

Condenser Focus Knob moves the condenser up or down to control the lighting focus on the specimen.

Exercise - 5

Viable Count of Bacterial by Serial Dilution Pour Plate Method

Microorganisms govern most of the soil fertility processes. Viable counts, or biomass in soil are generally correlated with the fertility level of soil and more so with the organic matter content. Hence, viable counts serve as an index of soil productivity. This procedure counts only a small portion of the total population but is useful for studying the relative abundance of soil bacteria and changes in density of that population that develops on the medium. The method involves the dispersion of the soil in the medium to such an extent that an individual cell, spore, or mycelial fragment has reasonable opportunity, when exposed to suitable conditions, to develop into a discrete and macroscopically visible colony. The necessary degree of dispersion is usually achieved by making successive dilutions of the given soil sample. Various factors that determine the accuracy of the plating method are:

1. The use of an appropriate medium. It is particularly important that the agar medium be low in readily available energy materials such as sugar or proteins in order that a few fast growing bacteria do not dominate growth on the plate and inhibit the development of a much larger number of slow growing species
2. Carefully controlled temperature of incubation
3. Length of incubation.
4. Error in sampling and in sample preparations.
5. The manner in which the colonies are counted.

Merits

1. Counts living population, which gives an idea of activity of different microbes
2. Any organism of interest can be isolated.

. Chromogenic nature of the organism in particular medium can be studied.

Demerits

. Every medium does not support the growth of all soil microbes.

. Some of the cell may stick to pipettes.

. While making dilution some cells might not get chance to come in suspension because of their nature of binding with soil clay. Some cells may rupture by mechanical shaking.

. Some inactive form of cell like endospore may also germinate in soil. Thus it is not totally correct to say that it counts active form of bacteria.

Materials

1. Finely sieved soil
2. Sterile water blanks 9 ml
3. Sterile pipettes 1 ml
4. Sterile petridishes
5. Colony counter
6. Microscope and slides
7. Medium: Soil extract Agar (Prepared in Exercise-I)

Procedure

1. Determine the moisture percentage of soil by drying 10-g sample to constant weight in an oven at 105°C.
2. Weigh and suspend 10 g of moist of soil in 90 ml water blank to obtain a 1/10 or 10^{-1} dilution. Transfer 1 ml of suspension to 9 ml water blank, it establishes a 10^{-2} dilution. From this second dilution, similarly transfer 1 ml quantities to water blank of 9 ml to provide dilution series ranging from 10^{-3}, 10^{-5}, 10^{-6} and 10^{-7} dilutions.
3. From the highest dilution prepared (assume 10^{-7}) transfer 1 ml portion of the freshly agitated suspension to each 3 sterile petridishes by means of sterile 1 ml pipette. Make similar transfers from the next lower dilutions (10^{-6}, 10^{-5} and 10^{-4}) into other dishes. Inoculate 1 ml water blank in 3 petri dishes. It will serve as control.
4. Into each seeded petridish pour about 12 ml of soil extract agar, which previously

has been steamed sufficiently to ensure complete melting and then cooled to 42 °C

5. Immediately after adding agar, rotate the dish by hand to distribute the so inoculums throughout the medium.
6. Permit the agar to solidify. Then invert them, and place them in an incubat at 28 °C for one week.
7. Upon removal from the incubator, inspect all plates prepared from a sing dilution to see whether a proper dilution has been plated and whether a prop dilution effect is apparent. The latter means that a plate from a 10^{-6} dilutio should have only approximately 1/10th as many colonies as the plate prepare from the 10-5 dilution. If there are numerous colonies on the plates, and if dilution effect is not apparent, contamination has undoubtedly occurred, and a plates should be discarded. The counts in control shall confirm contaminatio
8. If the plates from highest dilution show more than 300 colonies, the dilution h been too low, and if those from the lowest dilution show less than 30 coloni the dilutions has been too high. In either case, discard all plates. Select the plat from the dilution at which 30 to 300 colonies have developed. In one or tw plates within the group have one or more large bacterial or fungal colonies (cm in diameter) discard such plates without counting.
9. Calculate the results by multiplying average number of colonies per plate fc a given dilution by the dilution factor to obtain total viable count per gram c the initial moist soil sample.
10. Divide the count per g of moist soil by means of dry matter per gram of mois soil to convert the results on the dry soil basis.
11. The isolated bacteria are purified on Nutrient agar media slants and microscopi observations are made on cell morphology viz. shape, size and arrangemen following Gram staining and Negative staining.

Staining of bacteria

a. Simple staining: Negative staining or Relief staining

Nigrosin Stain-------------------------------------10 g

Water--100 ml

Boil for 30 minutes and add 0.5 ml formalin. Filter through filter paper and store i 2 ml amounts in sterile tubes. Place a loopful of the suspension to be examined o a slide. Immediately add an equal amount of the nigrosin solution, mix and sprea out in a thin film.

Differential staining

ram staining

ram differential stain was developed by Christian Gram (1884). The stains quired for Gram staining are:

Crystal violet

olution A- 2g Crystal violet in 20 ml 95% alcohol
olution B- 0.8 g Ammonium oxalate in 80 ml distilled water.

Iix solutions A and B

Gram's Iodine

odine crystals – 1 g
otassium iodide – 2 g
istilled water – 300 ml

llow to stand 24 hours for the Iodine to dissolve. It may be necessary to add a few ore crystals of Potassium iodide.

. **Ethyl alcohol 95%**

. **Saffranine**

.5 per cent solution in 95% alcohol-----------25 ml
istilled water-------------------------------------75 ml

rocedure

Prepare homogenous bacterial smears from one-day-old broth culture on clean glass slides, air dry and heat fix.

. Cover the smear with crystal violet stain for1 minute, wash with distilled water, flood with mordant Gram's Iodine solution for 1 minute. Wash the slides with distilled water.

i. Decolourise with absolute alcohol until no violet colour comes off.

. Counter stain the smear with safranine for 30 seconds.

Wash the smear, blot dry and observe under microscope using oil immersion objective.

vi. Gram-positive bacteria will appear bluish purple while the Gram-negati pinkish red in colour.

Motility Test (Hanging Drop Technique)

1. Apply petroleum jelly to the edges of a cleaned cover slip.
2. Transfer a loopful of 1-day-old culture suspension in broth to the cen of the cover slip.
3. Place a cavity slide over the cover slip, with the cavity facing downwar covering the suspension.
4. Press the slide to seal up the cover slip and turn over quickly.
5. Observe the preparation under the low and high power objectives for testi the motility of bacteria.
6. The bacteria that move swiftly across the microscopic field are connot positive for motility.

Observations and Calculations

No. of bacterial colonies in different dilutions				
Plate (Replication)	10^{-4}	10^{-5}	10^{-6}	10^{-7}
1				
2				
3				
4				
Average				
	$\times 10^4$	$\times 10^5$	$\times 10^6$	$\times 10^7$

Select the dilution in which the numbers of colonies fall between 30-300. Reje the other dilutions.

No. of viable bacteria per gram of wet soil $= X \times$ Selected dilution

No. of viable bacteria per gram of dry soil $= \frac{X \times 100}{90}$
(Assuming 10% moisture on wet weight basis)

References

Gerhardt, P., R.G.E. Murray, R.N. Costilow, E.W. Nester, W.A. Wood, N.R. Kreig and G. B. Phillips (1981) Manual of Methods for General Bacteriology, Washington: American Society for Microbiology.

Murray, R.G.E. and C.F. Robinow (1981). Manual of Methods for General Bacteriology, Washington.

Starr, M. P., H. Stolp, h. G. Truper, A. Balows, and H. G. Schlegel (ed.) (1981). The Prokaryotes. A handbook on habitats, isolation and identification of bacteria. Vol. II. Springer-Verlag New York, New York.

Questions

1. Why soil extract medium is used for viable counts of bacteria?
2. Why media with low energy source is used in this experiment?
3. Which portion of the bacteria in soil is counted by this method?
4. Soil extract agar is of which type media?
5. List factors, which determine the accuracy of plating method.
6. Why media is cooled to about 42 °C for plating?
7. What is the importance of viable counts of bacteria in soil?
8. What is the purpose of heating dried bacterial smears before staining with dye?
9. Why is Gram stain of greater value than a simple stain?
10. Why do some organisms stain Gram positive and others Gram negative?
11. What purpose does alcohol serve?
12. What is a mordant?
13. At what age a bacterial culture is actively mobile?

Exercise - 6

Viable Count of Fungal in Soil by Serial Dilution Pour Plate Method

Fungi constitute an important part of micro flora of normal soil. In soils fungal propagules vary from 2×10^4 to 1×10^6 per gram soil. The propagule being considered as fungal hyphae, spore, and hyphal fragments which is capable of giving colony on the plate. Most soil forms have a filamentous organisation and are capable for reproduction by means of spores. Soil fungi has significant role in organic matter decomposition, as a plant pathogens and in the interaction with other soil microorganisms. They are first to attack the straw added in soil. Agar plate count method is commonly used as a measure of fungal population in soil, although there are few limitations of this method.

1. The method fails to distinguish Fungi constitute an important part of the microflora of normal soil. In soils, fungal propagules vary from 2×10^4 to 1×10^6 per gram of soil. The propagule being considered is fungal hyphae spores, and hyphal fragments, which are capable of giving colonies on the plate. Most soil forms have a filamentous organization and are capable of reproduction by means of spores. Soil fungi have a significant role in organic matter decomposition, as plant pathogens, and in the interaction with other soil microorganisms. They are first to attack the straw added in soil. The agar plate count method is commonly used as a measure of fungal population in soil, although there are a few limitations to this method. distinguish between spores and hyphae as a source of colonies.
2. The method is biased strongly toward organisms that grow rapidly and sporulate rapidly.
3. The mere act of shaking often introduces an error into the population estimates; the agitation tends to rupture the mycelium and sporulating body into an indeterminate number of fragments each of which may produce a single colony. So this always gives more count than actually found in soil.

There is nevertheless a distinct place for plating technique since it not only shows relative numbers of viable units under different management practices but also serves as one of several qualitative means of demonstrating the types of fungi and characteristics of a given habitat. Various media have been employed. The growth of bacteria and actinomycetes, which are more abundant in soil, is suppressed by acidification of the plating medium to a pH of about 4.5 or by adding some specific inhibitors, which are antibiotics such as oxgall, sodium benzoate, rose bengal, streptomycin, and novobiocin. These treatments inhibit most streptomycetes as well as most bacteria from growing.

Materials

1. Soil sample
2. Sterile petridishes
3. Sterile pipettes
4. Water blank 95 and 9 ml
5. Colony counter
6. Microscope
7. Rose bengal streptomycin agar. Martin (1950) (prepared in Ist exercise)

Procedure

1. Find moisture content of the soil samples and prepare ten-fold dilution series ranging from 10^{-1} to 10^{-6} as described in Exercise II. For many soils satisfactory dilution range will be 10^{-3} to 10^{-6}.
2. Transfer 1 ml aliquot of each dilution to each 3 dishes after shaking the dilution by hand for a few seconds before withdrawing sample.
3. Transfer 1 ml aliquot from water blank into 3 petri dishes to serve as control.
4. Add approximately 12 ml of the medium, melted and cooled to approximately 42 °C.
5. Mix the soil suspension with the melted agar by rotating or swirling the freshly poured agar.
6. After the agar has solidified invert the plates and incubate them at 28 °C for 5 to 7 days.
7. Select for counting the dilution range that yielded from 10 to 100 colonies per plate. Include in the plate count only those colonies, which are typically filamentous and not bacterial in composition.

8. Convert the average count to a number per gram basis as described in Exercis II. The isolated fungal colonies are purified on Potato dextrose agar slants an are identified.
9. After incubation note the cultural characteristics of the organisms that develo Record differences in pigmentation, texture, and form of colonies.
10. Examine representative colonies with the low power objective of the microscop remove a small bit of the colony to a glass slide and examine in a drop of lacto phen mounting fluid under a cover slip. Make drawings of the different forms observe

Preparation of Microscopic Mounts and Staining of fungi:

The stains required for fungal staining are

1. **Lactophenol:** Mix 20 g Phenol (crystals), 20 ml lactic acid, 40 ml glyceri in 20 ml water.
2. **Lactophenol-Cotton Blue:** Combine 100 ml lactophenol, 5 ml aqueou solution of cotton blue, and 20 ml glacial acetic acid.

Procedure

» Prepare microscopic mounts in both 50% glycerol and lactophenol-cotton blu by taking a bit of culture growth with the help of sterile needle and forcep

» Place cover slips gently over the mounts and seal with D.P.X. mountant (mixture of Distyrene, a plasticizer and Xylene).

» Record microscopic features using high and low power and oil immersio objectives of the fungi.

Observations and Calculations

No. of colonies in each dilution				
Plate No.	10^{-3}	10^{-4}	10^{-5}	10^{-6}
1				
2				
3				
4				
5				
Average				
	$\times 10^3$	$\times 10^4$	$\times 10^5$	$\times 10^6$

elect the dilution where the number f colonies are in between 10-100 colonies er plate.

ungus propagules per gram of wet soil = X

ungus propagules per gram in dry soil = $\frac{X \times 100}{90}$

Assuming 10% moisture on wet weight basis)

eferences

arley, J. L. and Waid, J. W. (1955) A method for studying active mycelia on living roots and other surface in the soil. *Trans Br. Mycol. Soc.* 38: 104-118.

anwar, S.S., Gupta, M.K. and Punj, V. (1997) Laboratory Manual of Basic Microbiology, Department of Microbiology, College of Basic Sciences, Himachal Pradesh Krishi Vishvavidyala, Palampur.

uestions

What is the role of fungi in soil?

Why streptomycin is added to Rose Bengal streptomycin agar?

What are the drawbacks of plate count method for counting fungi? Which plates should be selected for counting fungi by agar plate method?

Name inhibitors, which can allow the fungus to grow and not other organisms?

The fungal colonies developed from spores are counted in this method, comment.

Exercise - 7

Viable Count of Actinomycete in Soil by Serial Dilution Pour Plate Method

The actinomycetes, sometimes called ray fungi, are intermediate in morpholo between the bacteria and the true fungi. They possess a very fine vegetative myceliu that does not fragment into bacillary or coccoid forms, and therefore their colonies agar are commonly cohesive or though in consistency. In many soils, the actinomyce account for from 10 to 25% of the total count on agar plates and in soils for mc than 50%. Their number is higher in soils of low moisture content than in wet soi They are responsible for degrading more resistant carbon material in soil like hum They are also important flora of manure heaps.

The counting of actinomycetes populations of soil by plate count method complicated because no cultural substrate or medium has been developed that highly selective for actinomycetes. The plates commonly contain colonies of fun and bacteria along with those of actinomycetes. For accurate counts of actinomycet the colonies are first differentiated from bacterial and fungal colonies, and then th are counted. Plate count of actinomycetes has similar limitations as that of fung count. Hence their counts by this method are generally high.

Materials

1. Soil sample
2. Sterile petridishes, pipettes
3. Water blanks 95 ml and 9 ml
4. Dextrose nitrate agar medium (Ken-knight and Muncie, 1939) (Prepared Exercise I).

Procedure

1. Determine the moisture content of the soil by drying 10 g of soil sample to constant weight in an oven at 105 °C.
2. Prepare serial dilution from 10^{-1} to 10^{-6} (As per Exercise III).
3. Pipette out 1-ml aliquot from selected dilution in 5 plates for each dilution. Maintain control.
4. Pour approximately 12 ml of dextrose nitrate agar to each plate.
5. Rotate the petriplates to distribute the soil inoculum throughout the medium. Permit the agar to solidify, invert and incubate the plates at 28 °C for 7 days.
6. At the end of the incubation period, select the plates for counting, showing at least 30 and not more than 300 colonies per plate. If in doubt in identifying a colony as bacterial or actinomycetal, probe the colony with a dissecting needle. If the colony tends to remain as a discrete small mass when probed, it is probably an actinomycete. If a colony tends to smear out and lose its sharp periphery on probing, then it is probably a bacterium. If a colony is doubtful prepare a stained smear.
7. For the average colony count, calculate the population per g of soil as described.

Observations and Calculations

Number of colonies in each dilution				
Plate No.	10^{-3}	10^{-4}	10^{-5}	10^{-6}
1				
2				
3				
4				
Average				
	$\times 10^{-3}$	$\times 10^{-4}$	$\times 10^{-5}$	$\times 10^{-6}$

Select dilution-having colonies between 30-300.

Actinomycete population per gram of wet soil $= X$

Actinomycete population per Gram of dry soil $= \frac{X \times 100}{90}$

(Assuming 10% moisture on wet weight basis)

Reference

Nonomura, H., and Y. Ohara (1969). Distribution of actinomycetes in soil. IV A culture method effective for both preferential isolation and enumeratio of *Microbispora* and *Streptosporangium* strains in soi (Part I). Hakku Kogak Zasshi 49: 895-903.

Questions

1. How you will identify the actinomycetal colony?
2. Why fungus and bacteria grow on this media of actinomycetes?
3. Differentiate between inoculation and incubation.
4. With which number of actinomycetes colonies the plates are selected fo recording results?
5. Actinomycetes are the transition between true bacteria and fungi. Comment.

Exercise - 8

Preparation and Sterilization of Media and other Glasswares

Selection of a medium to grow any particular type of microbe is extremely important since the microbes vary in their nutritional requirements. Certain nutrients are needed by almost all the microbes and are called basal medium. Each genus has some specific nutritional requirements. Thus, different media are used for different groups of microbes. The following are the main types of media that will be used in this course:

a. General purpose media (Nutrient agar)
b. Selective media (N-free medium for free living N_2-fixers)
c. Differential media

Procedure

Prepare the following the media namely Nutrient Agar/Soil Extract Agar, Dextrose Nitrate Agar, Potato Dextrose Agar/Rose Bengal Streptomycin Agar (Martin, 1950)/ Czapek-Dox Agar for bacteria, actinomycetes and fungi, respectively. Composition of these media is as follows:

a. Bacteria

a) Nutrient agar	
Peptone	5 g
Beef extract	3 g
Sodium chloride	5 g
Distilled water	1000 ml
pH	6.8-7.0

Agar	15 g
Cycloheximide	40 mg/l of autoclaved medium

b) Soil extract agar	
Glucose	1.0 g
Dipotassium hydrogen phosphate	0.5 g
Potassium nitrate	0.1 g
Soil extract*	100 ml
Distilled water	900 ml
pH	6.5 - 7.0
Agar	15 g

*Soil extract is prepared by mixing one kg of fertile soil and 0.5g $CaCO_3$ with 100 ml of tap water and heating in autoclave for 20 min. $CaCO_3$ flocculates colloid material and clarify. Filter extract and make the volume to one litre.

b. Actinomycetes

a) Dextrose nitrate agar	
Dextrose	1.0 g
Potassium dihydrogen phosphate	0.1 g
Sodium nitrate	0.1 g
Potassium chloride	0.1 g
Magnesium sulphate	0.1 g
Distilled water	1000 ml
pH	6.8-7.0
Agar	15 g
Cycloheximide	25 mg/l of autoclaved medium
Amphotericin B	10 mg/l of autoclaved medium
Novobiocin	25 mg/l of autoclaved medium

c. Fungi

a) Potato Dextrose Agar	
Potato (Peeled)	200 g
Dextrose	20.0 g

Agar	15.0 g
Distilled water	1000 ml
Streptomycin	75 mg/l of autoclaved medium

ut potatoes into small pieces and boil them in one litre of distilled water for 20 inutes. Strain the broth through 3-4 layers of cheesecloth and discard the potato eces. To this broth add dextrose, agar (with stirring) and water to make volume 1 litre. Autoclave it.

b) Peptone Rose Bengal streptomycin agar (Martin, 1950)	
Dextrose	10.0 g
Peptone	5.0 g
Potassium di-hydrogen Phosphate	1.0 g
Magnesium sulphate	0.5 g
Agar	15.0 g
Rose Bengal	0.033 g
Distilled water	1000 ml
Streptomycin**	75 mg/l of autoclaved medium

In case of rose Bengal streptomycin-agar, sterilized streptomycin solution (by ltration) be added @ 75 μg/ml of the autoclaved medium.

:) Czapek-Dox agar medium	
Sodium nitrate	2.0 g
Dipotassium hydrogen orthophosphate	1.0 g
Magnesium sulphate	0.5 g
Potassium chloride	0.5 g
Ferrous sulphate	0.01 g
Sucrose	30 g
Agar	20 g
Distilled water	1000 ml
pH	6.8-7
Streptomycin**	75 mg/l of autoclaved medium

Functions of ingredients

For hetrotrophic soil microorganisms organic source of carbon is needed which provided through glucose, dextrose, peeled potatoes, beef extract and mannitol. Sor salts like KNO_3, $NaNO_3$, NaCl, $MgSO_4$, KCl and soil extract are added to provi nutrients. The use of K_2HPO_4 and KH_2PO_4 besides providing K and P also he in maintaining the pH of the medium by acting as buffer. Peptone and beef extra provide the organic source of N and vitamin B and malt extracts provide growt promoting substances like vitamins. Streptomycin sulphate in Potato dextrose aga Peptone Rose Bengal streptomycin agar acts as an antibacterial agent. The ag solidifying agent is an extract of seaweed marine algae viz. *Geledium*, *Gracilari* etc. Agar will harden a solution into a clear jell similar to ordinary gelatin, but it different from gelatin in certain respects that make it much more suitable for use the microbiology laboratory. For example, it is non-nutritive for most organism whereas many organisms utilized gelatin (a protein) as a nutrient and as food. It converted into liquid by the microorganism and the solidifying properties that originally had are destroyed. Agar usually is not utilized as food and will remain sol Another advantage of agar results from its melting and solidifying temperatur Incubation of most microorganisms is conducted at the melting temperature gelatin (25°C). A medium solidified by gelatin will therefore become fluid. Ag liquifies beyond 100°C but once it has molten it will remain fluid to approximate 40°C, just a little warmer than your body temperature. Both agar and gelatin for a clear transparent material when hard, thus making it possible to see through reasonably thin layer of this medium.

All the ingredients of these media are weighed and dissolved in small amou of distilled water except agar-agar. Make up the volume. Adjust the pH of the med with dilute HCl or dilute NaOH. Transfer the media in a pan, now add agar-ag (nonabsorbent) and boil with occasional stirring. Pour the media in flasks or tub and plug them with cotton and sterilize.

Preparation of common microbiological apparatus

The apparatus generally used for microbiological purpose are, pipettes, petri dish sterile water blanks and glass slides etc.

a. **Pipettes:** Pipettes are used for making dilutions, transferring solutions ar cultures. Thoroughly wash and dry the pipette, plug with a cotton plug the flat end and place-tapered end downwards in metal container. Fill th metal container with pipettes and cover with slip lid. They are now reac for sterilization in an oven.

b. **Petri-dishes**: These are used to grow the microorganism. Washed and properly assembled petri dishes are wrapped in paper singly or in pairs and sterilized.

c. **Sterile water blanks:** These are used for making dilutions, or can also serve as a control treatment in some studies. Dispense the desired amount of distilled water in test tubes or conical flasks. Plug them with cotton (cap the screw cap tubes loosely) and sterilize.

Sterilization

Microbiologically sterilization means making media free of all the forms of life. Different materials are sterilized differently. The choice of the method depends upon the nature of the material to be sterilized. The following are various methods of sterilization.

a. Dry heat sterilization (hot air, sterilization)
b. Wet heat sterilization (steam sterilization)
c. Chemical sterilization
d. Flame sterilization
e. Filter sterilization
f. Radiation method

a. **Dry heat:** Dry heat means oven heat. Materials to be sterilized are kept in an oven at 160°C for 2 hours. The glasswares, which are not destroyed by high temperature are sterilized by this method, e.g. pipette, petri dishes, etc. After sterilization they are stored, while still wrapped in paper, the paper cover removed just before using.

b. **Wet heat**: Sterilization by this method is done where we want to retain the moisture and low temperature to prevent hydrolysis and chemical reactions in the material. The equipment used is called autoclave, in which steam heat under pressure is effective force in killing the microbes. The normally required media are sterilized by it at 15 Ib sq inch^{-2} pressure (121.6°C) for 15 to 20 min. while soil is sterilized for 1 hr at 15 Ib sq inch^{-2} pressure for three consecutive days. Sterilization of syringes etc. is obtained by boiling water at 100°C.

c. **Chemical sterilization:** The chemicals such as phenyl mercuric acid, methyl bromide, alcohol, and mercuric chloride are the effective killing agents. Some materials are treated with these chemicals to sterilize them. Nodules are surface sterilized by dipping in 0.1% solution of mercuric chloride and in 70% alcoho

d. **Flame sterilization**: Some of the materials are sterilized by making them red ho on the flame. The inoculation needles, forceps etc. are sterilized by this methoc

e. **Filtration:** Only liquid solution, which are heat labile are sterilized by thi method. The solution to be sterilized is passed through a Millipore filter pa of a particular pore size (< 0.045µ) to get rid of microbial bodies. Gentle ai suction is generally required to obtain a sterile solution. Pesticides, hormones radioactive chemicals, blood plasma, etc. are sterilized by this method.

f. **Radiation:** This is one of the mildest ways of sterilization, employed for solids liquids or gases where high temperature is undesirable. Tabletops, chambers etc can be exposed to U.V. light (200 to 320 mµ wave length) for ½ hour. Irradiatio with X-rays from cobalt-60 is used with great advantages. The normally presen ultra violet rays in the sunlight also have a mild sterilization effect.

Reference

Wollum II A G (1982) Cultural methods for soil microorganisms. In A L Page, F H Miller and D R. Keeney (ed.). Methods of Soil Analysis, Part II. Chemica and Microbiological properties. Agron. Monograph. 9 (Second Ed): 781-802

Questions

1. Why soil is autoclaved for three consecutive days for its sterilization?
2. Write short note on various methods of sterilization.
3. Why do you use three different media for growing bacteria, fungi anc actinomycetes?
4. Write down the purpose of adding various ingredients in the soil extract aga medium.
5. Why agar-agar is used as a solidifying agent in the medium?
6. 6. What is the function of streptomycin, cycloheximide, amphotericin B anc novobiocin in different media?

Exercise - 9

Study and Isolation of *Rhizobium* from the Nodules

Rhizobium is heterotrophic soil bacteria that can fix appreciable amount of atmospheric nitrogen, in association with the compatible host legume. This type of association where both the partners (Symbiont) are benefited is termed as symbiosis. A group of host plants infected by a single species of *Rhizobium* is called inoculation group. There are about 20 cross inoculation groups of legumes but only seven are important.

Infection is the ability of a strain to nodulate a given host. The capacity of nodule to bring about N_2-fixation in conjunction with the plant varies greatly. The relative capacity of symbiotic system to assimilate molecular nitrogen is known as its effectiveness. Some *Rhizobium* spp. are highly effective, while others are completely ineffective in nitrogen fixation. The mere presence of nodules on a legume plant not necessarily benefits the plants. Given below are the ranges of N_2fixed by different crops.

Crop	Kg N/ha/annum	Crop	Kg N/ha/annum
Clover	50-200	Soybean	40-250
Alfalfa	40-350	Pea	30-140
Cowpea	20-250	Beans	40-150

The various parameters to judge the effectiveness of a strain are size, colour and site of nodule on root, N content, yield of plant and N_2-ase activity. When the *Rhizobium* cells multiply in nodule, they transform into Bacterioids. These are the physiologically active, lacking in cell wall, irregular cell. However, nodule contains some unchanged rod of rhizobia that can grow on petri-plate in presence of suitable carbon source. Thus we are able to isolate *Rhizobium* from nodules.

Materials

1. Big size healthy nodules
2. Microscopic slides
3. Rose Bengal stain
4. Forceps
5. Razor blade
6. Mercuric chloride (0.1%)
7. 95% Ethyl alcohol
8. Sterile Petri dishes
9. Congo red mannitol agar

Selective media- Congored Manitol agar

Mannitol	10.0 g
K_2HPO_4	0.5 g
$MgSO_4.7H_2O$	0.2 g
NaCl	0.1 g
$CaCO_3$	3.0 g
Yeast extract	1.0 g
Agar	10.0 g
Congored	5 ml (.5% solution)
Distilled water	1 litre

Procedure

a. To examine percent effective nodules

1. Wash roots free of adhering dirt
2. Cut 10 nodules randomly and note the colour of the juice coming out from nodules on crushing; if pink in colour it means it is effective, otherwise not
3. Record the percentage of effective and ineffective nodules
4. Record percentage of nodules on main and lateral roots
5. Note the colour, size, shape, abundance and location of the characteristic nodules. Make appropriate notebook sketches.

Examination of nodules for Bacteroids

1. Sterilise the nodules with 0.1%-acidified mercuric chloride for 2 minutes and wash in sterile water.
2. Crush the nodule in a drop of water on a clean slide; mix well with a loop.
3. Transfer several loop-full to another clean slide
4. Dry and fix the slide on gentle heat and stain with rose bengal stain for 5 minutes.
5. Identify bacteroids having irregular shapes and an uneven distribution of protoplasmic contents.

Isolation of *Rhizobium* from nodules

1. Wash the soil from the root nodules and carefully out a nodule from the root, leaving a small portion of the root attached.
2. Place the nodule in a petri plate containing 0.1% acidified $HgCl_2$ (1 ml conc. HCl/1) and keep it immersed for 5 minutes.
3. Use sterile forceps to transfer the nodule to a petri plate containing sterile water. Wash at least 6 times, and place in 95% ethyl alcohol for 3 minutes. Remove to a petri plate containing sterile water and rinse the nodule twice.
4. Add 1.0 ml of sterile water to each of six numbered sterile petri plates. Crush the thus sterilised nodule with forceps in petri plate No. 1. Mix the nodular tissue with water.
5. Transfer one loop full of the suspension to plate 2 and dilute by mixing with 1.0 ml of water. Repeat the loop dilution procedure for plates No.3, 4, 5 and 6.
6. Pour plates (NO. 2-6) with congo red mannitol agar and incubate the plates at 28 °C for 5-7 days.
7. After incubation examine the plates, note typical *Rhizobium* colonies; raised, moist, with round edges, later changing to opaque white. Stain with carbolerythrosin (5 minutes) and compare the bacteria with those found in the nodule.

ultural morphological and staining characters of *rhizobium*

he isolation of *Rhizobium* from legume nodules not necessarily result in rhizobia ecause bacteria of related genus *Agrobacterium* may also grow on agar plates along ith *Rhizobium*. The sure way of identifying *Rhizobia* is to conduct plant infection sts and see if effective nodules are formed. A plant test is done under bacteriological controlled conditions with assemblies ranging from small tubes to pots. Befc these tests are done for a large number of cultures one may have to carryout certa routine following cultural tests.

Growth on solid media

Rhizobium grows very poorly on peptone glucose agar but grow well on yea mannitol agar in the form of watery or white colonies. Most strains produce gu of varying composition.

Congo red test

This test is designed to differentiate *Rhizobium* from *Agrobacterium*. *Rhizobiu* grows as white, translucent, glistening, elevated and comparatively small coloni with entire margins in contrast to stained colonies of *Agrobacterium*, on yeast extra mannitol agar containing 2.5 ml of 1-% congo red solution.

Hofers alkaline broth test

The test is based on the fact that *Agrobacteria* grow at higher pH levels, while *Rhizob* are unable to do so. A medium heaving pH 11.0 is used to screen new isolates.

Lactose test

*Agrobacteri*a utilise lactose by the action of enzyme ketolactase, whereas *Rhizob* cannot utilise the sugar. This can be detected on agar medium containing lacto at 10 g/1. Nodule bacteria are grown on lactose agar plates for 4-10 days and th plates are flooded with Benedict's reagent. The formation of yellow colouration du to Cu_2O indicates the presence of *Agrobacteria*.

Microscopy

Under a phase contrast microscope, *Rhizobia* are visible even in simple water moun and show up the presence of characteristics refractive granules of ß-hydroxy butyra within cells. The characteristic ß-hydroxy butyrate granules can also be seen und bright field microscope after staining with diluted carbol fuchsin. *Rhizobium* cel are small to medium sized rods (0.5-0.9 x 1.2x3.0 μ). They are motile when your and have bipolar, subpolar or peritrichous flagella. The *Rhizobium* cells have a larg irregular shaped nuclear region in the centre surrounded by a narrow region o denser protoplasm. Rhizobia are Gram negative (Study Gram reaction by stainin as in Exercise II.

Observations and Calculations

1. Effective nodules out of ten = X10
2. Percentage of effective nodules = X/10 x 100 = 100
3. Draw the shapes of bacteriod cells
4. Are you able to isolate *Rhizobium* for nodules?

Reference

Fred, E. B., Baldwin, J.L. and McCoy, E. (1932). Root nodule bacteria and leguminous plants. University of Wisconsin Studies, Madison Science, 5: 343-349.

Subba Rao, N.S. (1993). Biofertilizers in agriculture and forestry. Oxford and IBH Publishing Co. Pvt. Ltd., Pg. 70-73.

Questions

1. What is a bacteroid?
2. Differentiate between ineffectiveness and effectiveness?
3. From which cells *Rhizobium* colony originates while isolating *Rhizobium* from nodules?
4. What is symbiosis?
5. What is a cross inoculation group?
6. Why the nodules are surface sterilised for isolating *Rhizobium*?

Exercise - 10

Azotobacter Population in Soil by Serial Dilution Pour Plate Method

Introduction

The fixation of atmospheric nitrogen by free-living microorganisms without another host system is known as non-symbiotic nitrogen fixation. It is restricted to certain microorganisms, mostly bacterial and blue-green algae. The bacteria that have the ability to fix molecular nitrogen come under the genera Azotobacter, Azomonas, Beijerinckia, Derixa, Mycobacteria, Azospirillum, Bacillus, Enterobacter, Klebsiella, and Clostridium. Different strains of Azotobacter are able to fix 2-15 mg N/g of carbon source utilized. Although higher values have been often reported. Beijerink was the first to isolate and describe Azotobacter. The prominent species of Azotobacter recognized are as follows: chroococcum, agilis, vinelandii, beijerinckii, insignis, microcytogenes, and paspali.

The A. chroococcum is the predominant species of Azotobacter in Indian soils, and the population rarely exceeds 10^4 to 10^5 cells/g of soil. Azotobacter is an aerobic and heterotrophic organism. Therefore, depend upon energy derived from degradation of plant residues. There are two factors that have a major influence on the population of Azotobacter in soil. They are the associative and antagonistic action of soil microflora and organic matter content of soil. Soil pH, moisture, cultivation and temperature affect the occurrence of Azotobacter in soil. Azotobacter is absent or is present in a few numbers in the environments more acidic than pH 6.0.

Generally, Azotobacter spp. can be isolated and counted from soil using the dilution pour plate method. After incubation of three days, Azotobacter colonies develop on the agar medium. The MPN method has also been used.

In general, Azotobacter cells are pleomorphic; the size of young rods varies from 2.0 x 1.0 to 2.5 μ, and occasionally an adult cell may measure 10-12 μ. The morphology f the cells, however, is dependent upon the composition of the medium. Young cells ave peritrichous flagella, which serve as locomotory organs. A. chroococcum, A. inelandii, and A. beijerinckii are related species, and their cells are generally rod-haped with variable cell size. On the other hand, in A. agilis and A. insignis, the ize of the cell and its oval to coccoid shape undergo little change due to aging or utritional conditions. A. chroococcum, A. vinelandii, A. beijerinckii, and A. paspali re known to form cysts to withstand adverse conditions. Each cyst has a living cell vith two coats. The cyst accumulates poly-β-hydroxybutyric acid. Polysaccharide or um production is another characteristic feature of A. chroococcum, A. beijerinckiii, nd A. vinelandii; on the other hand, A. agilis produces very little gum.

A chroococcum produces characteristic black pigment, melanin, especially in lder cultures. This pigmentation is due to the oxidation of tyrosine by tyrosinase, n enzyme that has copper. The formation of yellow pigment is characteristic of A. eijerinkii. A. vinelandii produces green-yellow water-soluble fluorescent pigment vhile older cultures of A. insignis are yellow-brown in color. Azotobacter cells are ram negative as they retain the color of secondary dye.

Materials

1. Jensen's medium
2. Sterile Petriplates
3. Sterile pipette
4. Water Blank, nine ml
5. Soil sample
6. Glass sticks
7. Slides
8. Stains for Gram reaction (Exercise II)
9. Microscopes

Jensen's medium

Sucrose	20.0 g
K_2HPO_4	1.0 g
$MgSO_4 7H_2O$	0.5 g
NaCl	0.5 g

$CaCO_3$	0.5 g
$FeCl_3$	0.001 g
$NaMoO_4$	0.001 g
D.W.	1 litre
pH	7.5
Agar	20.0 g

Procedure

1. Determine the moisture content of the given soil sample
2. Prepare serial dilutions of soil upto 10^{-4}
3. Pour the Jensen's medium in petridishes and let it solidify. It will be better if th medium is added in petridishes a day before the actual inoculation.
4. Inoculate 0.1 ml of each suspension on the surface of solidified medium. A se of three petridishes should be inoculated from each dilution. Three petridishe are kept as control.
5. Spread the inoculum on the surface of the medium with the help of sterile glas slide. The glass stick can be sterilised by dipping in the alcohol and flaming.
6. Petridishes are incubated in inverted position at 28 °C for 3-5 days.
7. The soft, milky and mucoid colonies of *Azotobacter* can be very easily counte with naked eyes.
8. The results can be computed on dry weight basis of soil.

Cultural characters

The cultural characters of *Azotobacter* isolates on Jensen's can be noted as:

Form/shape	Circular
Elevation	Convex
Opacity	White, watery when young, black brown when old
Edge	Entire
Surface	Glistening

Staining and morphological characters

Stain the isolated culture for Gram reaction (procedure as in exercise II) ote the Gram reaction,

ell shape and size under oil immersion objective

Gram reaction	Gram negative
Cell shape	Young cultures rods and old culture pleomorphic
Cell size	2.0 2.7 x 1.0-2.5 µ old cells may measure up to 10 µ

bservations

Population

Colony counts on each petridish					
Replications	Control	10^{-1}	10^{-2}	10^{-3}	10^{-4}
1					
2					
3					
Total					
Mean					

umber of cells per *g* fresh soil = Mean number of colonies - Mean No. of colonies
ı control x dilution factor = Y

Ioisture pre cent of dry wt. basis =Number of cells per *g* dry soil = Y x moisture factor

. Cultural characters

Form/shape	
Elevation	
Opacity	
Edge	
Colour	
Surface	

. Morphological and Staining characters

Gram reaction	
Cell shape	
Cell size	

Reference

Beaucham, E.G. and Hume, D.J. (1997) Microorganisms to enhance nitrog fixation. In: Modern Soil Microbiology (eds.) Van Elasa, J.D., Trevres, J. Wellington, E.M.H. Pg. 658-659.

Questions

1. Which is the nitrogen source in Jensen's broth?
2. Which is the suitable range of pH for *Azotobacter* growth?
3. Discuss various methods used for isolation of *Azotobacter*.
4. Why the *Azotobacter* suspension is spread on the surface of solidified agar?
5. Describe the culture characters of *Azotobacter* on Jensen's agar
6. Describe the morphological characters of *Azotobacter* cells
7. Write about pigments produced by different species of *Azotobacter*
8. Name different species of *Azotobacter*

Exercise - 11

Population of Cyanobacteria (Blue Green Algae) in Soil and to Study their Morphological Characteristics

Soil inhabits many photosynthetic microorganisms. Of these prokaryotic blue-green algae, which contain phycocyanin pigment are also known as cyanobacteria. Cyanobacteria mostly prefer moist conditions, abundance of sunlight, mesophilic temperature and neutral to alkaline conditions. These organisms synthesize protoplasm and fix N_2 and can enhance the soil fertility. The contribution of soil algae (cyanobacteria, BGA) to agriculture has been established in rice (*Oryza sativa* L) fields.

Cyanobacteria such as *Anabaena, Nostoc, Tolypothrix* can compartmentalize their nitrogenase in specialized heterocysts in which a sufficiently reducing environment is maintained. They can fix N_2 under highly aerobic conditions. Other cyanobacteria e.g., the filamentous *Plectonema* and *Oscillatoria* and unicellular *Gloeocapsa*, do not produce heterocysts and generally synthesize nitrogenase under microaerophilic to anaerobic conditions. The population of Cyanobacteria can be determined by MPN method and plate count method by using any of following media. Such a count may include organisms that are unable to fix N_2 but are sufficient scavengers of fixed N.

Materials

1. Fresh soil (Take only from 0-2.5 cm top soil from moist area)
2. Microscope
3. Microscopic slides
4. Sterile dilution blanks
5. Sterile petri dishes

6. Sterile pipettes
7. Sterile culture tubes
8. Nitrogen and organic carbon-deficient medium for plate count

Medium

Fogg's Medium

A

Salt	Amount
KH_2PO_4	20 g/l
$MgSO_4.7H_2O$	20 g/l
$CaCl_2$	10 g/l
Iron Ethylene diamine tetra acetic acid (Fe EDTA)	1 g/25 ml

B

Salt	Amount
H_3BO_3	250 mg
$MnCl_2$	180 mg
$ZnSO_4$	20 mg
$CuSO_4$	5 mg
Distilled water	100 ml

Take 10 ml KH_2PO_4, 10ml $MgSO_4.7H_2O$, 10 ml $CaCl_2$, and 1 ml Iron EDTA from A and 1 ml of B in a volumetric flask and make volume to 1 litre. For pour plating add 15 g agar /litre. For MPN method pour 9 ml of medium broth in each tube. Autoclave the medium at15 lbs/square inch at 121°C for 15 minutes.

Procedure

Plate count method

1. Weigh 10 g soil and suspend in 90 ml of sterilised distilled water. Shake vigorously for 5 minutes and keep it as such for 10 minutes. This gives 10^{-1} dilution.
2. Prepare serial dilutions upto 10^{-4}.

3. Add 20 ml of melted medium in petri dish, swirl the petri plates and let the medium solidify.
4. Pour 0.1 ml suspension in each petri dish. Each dilution should be put in 3 plates.
5. Keep appropriate blanks also.
6. Incubate the petridishes under illumination of no more than 2500 lux from cool white fluorescent lamps with 16 h light alternating with 8 h dark at 30°C. The plates can also be put near window for light.
7. Count the colonies that develop after 4-6 weeks.

Most probable number method

1. Prepare serial dilution of soil as mentioned above.
2. From each of five successive serial dilution, usually 10^{-2} through 10^{-6}, transfer 1 ml to each of five incubation culture tubes containing about 9 ml sterile enrichment solution.
3. Keep the appropriate blanks also.
4. Incubate the tubes at 30 °C for 4-6 weeks in diffuse day light or under fluorescent lighting.
5. At weekly intervals during incubation, examine the inoculated tubes for green, blue green or other coloured growth such growths usually appear as surface rings or pellicles.
6. Record the number of tubes that are positive for algae at each dilution.
7. Find the most probable number of algae in soil as described for Nitrifiers in soil.

Microscopic examination

1. Transfer a portion of growth either from solidified medium or from the inoculated tubes to a microscopic slide by means of loop.
2. Make a direct microscopic examination of the wet mount using 10X and 40X objective lenses.
3. Note the size, shape and presence of heterocyst cells in the chain.

They are usually 10 μm or more in diameter, more commonly non-motile. Frequently appear in multi-cellular filaments of spherical or short cylindrical cells.

Observations

1. Calculate the population of blue green algae in the soil sample.
2. Draw the shape of algal cells as observed through microscope.
3. Count the number of heterocyst cells per filament.
4. Note the size of the cells.

References

Allen, O. N. (1957) Experiments in soil bacteriology. 3rd ed. Burgess Publishin Co., Minneapolis, USA.

Saxena, N.P. and Awasthi, D.K. (1985) Krishna's Microbiology. Krishna Prakasha Media (P) Ltd., Meerut. Pg. 141-178.

Questions

1. Why algae growth needs light during incubation?
2. What type of algae will flourish when Fogg's medium is used?
3. Differentiate the green and blue green algae.
4. What is heterocyst?
5. Why it is important to take surface 0-2.5 cm soil for study of algae population
6. What are microscopic features of algae examined in the class?

:xercise - 12

Study the Association between *Azolla* and *Anabaena Azollae*

zolla is a small water fern and is found floating on water in ponds and shallow tches. *Anabaena azollae* has been found to fix dinitrogen in cavities of its upper bes. *Azolla* has about 3% N content on dry weight basis and has long been used green manure in rice growing countries.

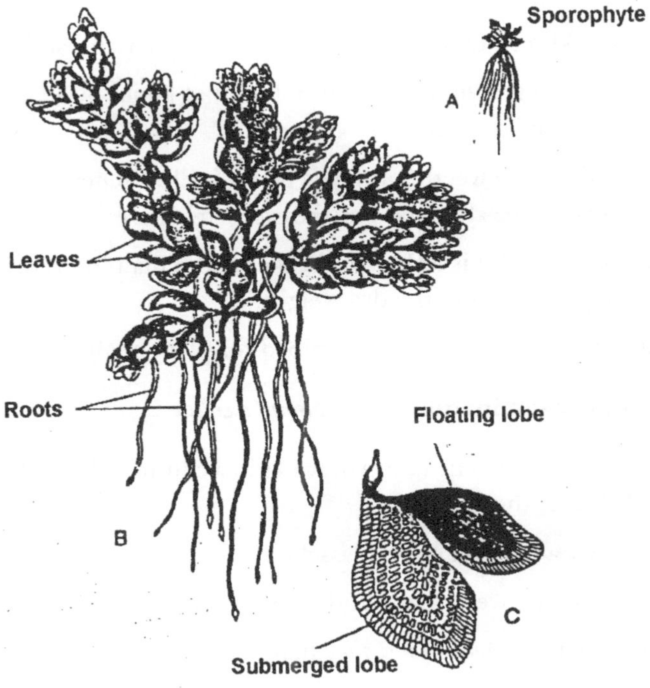

Materials

1. Culture tubes
2. Glass rods
3. Microscopic slides
4. Beakers 250 ml
5. Microscope
6. Sterile Petri dishes
7. *Azolla* fern
8. Incubating needle
9. Fogg's solidified and liquid medium (As given in Exercise on Algae)

Procedure

a) To study the growth and N_2 fixation capacity of *Azolla*:

1. Take 100 ml sterile medium in 250 ml beaker and add 1 g fresh *Azolla*.
2. Allow it to grown under diffused light at about 30°C or under fluoresce lighting for six weeks
3. Note the fresh weight of *Azolla* periodically after every week. The weighi should be done after draining the excess moisture for a constant period of ½
4. Inoculum of 1 g is left in the beaker and remaining is dried at 60 °C for overnight and note the dry weight.
5. Determine N content as described for plant analysis by Kjeldahl metho

b) Microscopic Examination of *Anabaena azollae*

1. Crush the *Azolla* in a drop of water in a culture tube with the help of a gla rod. Mix it thoroughly.
2. Transfer a loop on clean glass slide
3. Examine the wet mount under a microscope
4. Identify the *Anabaena azollae* filaments and see hetrocyst cells

c) Isolation of *Anabaena azollae*

1. Add 1 ml sterile water in sterile petridish.

2. Take a single *Azolla* plant and crush it with sterile glass rod and mix the content well.
3. Add about 15 ml melted, Fogg's medium and mix the contents thoroughly.
4. Incubate the plates under diffused light for 2-3 weeks at 30 °C.
5. Examine the growth under microscope and observe the cells of *Anabaena azollae*

Observations

A growth rate

Period (weeks)	Wet weight (g)	Dry weight (g)
0		
I		
II		

Microscopic

1. Cell shape
2. Heterocyst cell
3. Colour of the cells

References

Allen, O. N. (1957) Experiments in soil bacteriology. 3rd ed. Burgess Publishing Co., Minneapolis, USA

Somani, L.L. (2005) Handbook of Biofertilizers. Agrotech Publishing Academy, Udaipur. Pg. 43-178.

Questions

1. In which part of the fern the *Anabaena azollae* survive?
2. What are heterocysts?
3. What types of benefits are received by *Azolla* from *Anabaena* and vice-versa?

Glossary

Abiogenesis: Abio-means nonliving; genesis means origin.

Abiotic: Pertaining to or characterized by the absence of living organisms.

Acervulus: A sexual fruiting body or reproductive structure in a fungus.

Acid curd: Mill protein coagulated by acid.

Acid dye: A dye consisting of an acidic organic grouping of atoms (anion), which is the actively staining part, combined with a metal; the dye has affinity for cytoplasm.

Acid-fast: Retaining the initial stain and difficult to decolorize with acid alcohol. A property of certain bacteria.

Actinomycetes: Gram positive bacteria that are characterized by the formation of branching filaments.

Active transport: The energy requiring pumping of ions or other solutes across a cell membrane from a lower to a higher concentration.

Adaptive enzyme: An enzyme produced by an organism in response to the presence of the enzyme's substrate or a related substance. Also called induced enzyme.

Adenine: A purine component of nucleosides, nucleotides, and nucleic acids.

Adenosine triphosphate (ATP): A compound of one molecule each of a adenine and D-ri-bose and three molecules of phosphoric acid; it plays an important role in energy transformations in metabolism.

Adenoviruses: A group of icosahedral double-stranded DNA viruses.

Aerobe: An organism that requires oxygen for growth and can grow under an air atmosphere (21% oxygen). Compare anaerobe.

Aerosol: Atomized particles suspended in air.

Aflatoxin: The toxin produced by some strains of the fungus Aspergullus flavus; a carcinogen.

Agar: A dried polysaccharide extract of red algae (Rhodophyceae) used as a solidifying agent in microbiological media.

Agglutination: The clumping of cells.

Agglutinin: An antibody capable of causing the clumping or agglutination of bacteria or other cells.

Akinetes: Thick- walled single celled nonmotile asexual resting spores formed by the thickening of the parent cell wall; formed by some cyanobacteria.

Allergy: A type of antigen-antibody reaction marked by an exaggerated physiological response to a substance in sensitive individuals.

Allosteric enzymes: Regulatory enzymes with a binding or catalytic site for the substrate and a different site (the allosteric site) where a modulator acts.

Amino acid: An organic compound containing both amino (-NH2) and carboxyl (-COOH) groups.

Ammonification: The decomposition of organic nitrogen compounds, e.g., proteins, by microorganisms with the release of ammonia.

Amphitrichous: Having a single flagellum at each end of a cell.

Amylase: An enzyme that hydrolyzes starch.

Anabolism: The synthesis of cell constituents from simpler molecules, usually requiring energy. Compare catabolism.

Anaerobe: An organism that does not use O2 to obtain energy, cannot grow under an air atmosphere, and for which O2 is toxic. Compare aerobe.

Anaerobic respiration: Respiration under anaerobic conditions in which a terminal electron acceptor other than oxygen is involved.

Anaplasia: Structural abnormality in a cell or cells.

Antagonism: The killing, injury, or inhibition of growth of one species of microorganism by an other when one organism adversely affects the environment of the other.

Antheridium: pl. antheridia. A male gametangium.

Anthramycin: An antitumor antibiotic.

Antibiosis: An antagonistic association between two organisms in which one is adversely affected.

Antibiotic: A substance of microbial origin that has antimicrobial activity in very small amounts.

Antiobody: Any of a class of substances (proteins) produced by an animal in response to the introduction of an antigen.

Antigen: A substance that when introduced into an animal body stimulates th production of specific substance (antibodies) that react or unitie with th substance introduced (antigen).

Antimicrobial agent: Any chemical of biologic agent that either destroys or inhibi the growth of microorganisms.

Antiseptic: Acting against or opposing sepsis, putrefaction, or decay by eith preventing or arresting the growth of microorganisms.

Antiserum: Blood serum that contains antibodies.

Antitoxin: An antibody capable of uniting with and neutralizing a specific toxin.

Aperture: The magnitude of the angle subtended by the optical axis and the outermo rays still covered by the objective.

Aplanospore: A nonmotile spore; an abortive zoospore.

Apoenzyme: The protein moiety (portion) of an enzyme

Apothecium: A sexual fruiting body in a fungus.

Archaeobacteria/archaea: A major group of bacteria that includes the methanogen the red extreme halophiles, and the thermoacidophiles, and which diverge from other bacteria at a very early stage in evolution. Also called archaebacteri

Asepsis: A condition in which harmful microorganisms are absent, Adjective: asepti

Aseptic technique: Precautionary measures taken to prevent contamination.

Assay: The qualitative or quantitative determination of the components of a materia such as a drug.

Assimilation: The conversion of nutritive material into protoplasm.

Attenuation: A weakening; a reduction in virulence.

Autoclave: An apparatus using pressurized steam for sterilization.

Autolysis: The disintegration of cells by the action of their own enzymes.

Autotroph: A microorganism that uses inorganic materials as a source of nutrients carbon dioxide is the sole source carbon. Compared heterotroph.

Auxotrophic mutant: An organism having a growth requirement of specific nutrient not necessary in the parental strain.

Axenic culture: A microorganism of a single species, e.g., a bacterium, fungus, alga or protozoan, growting in a medium free of other living organisms.

Bacterial filter: A special type of filter through which bacterial cells cannot pass.

acterieide: An agent that destroys bacteria.

acteriochlorophyll: A chlorophyll like pigment possessed by anoxygenic photosynthetic bacteria.

acteriophage: A virus that infects bacteria and causes the lysis of bacterial cells.

acterium, pl. bacteria: Any of a group of diverse and ubiquitous prokaryotic single-celled microorganisms.

asic dye: A dye consisting of a basic organic grouping of atoms (cation), which is the actively staining part, combined with an acid, usually inorganic; the dye has affinity for nucleic acids.

ergey's Manual: An international reference work, which classifies and describes bacteria.

iochemical oxygen demand (BOD): A measure of the amount of oxygen consumed in biological processes that break down organic matter in water; a measure of the organic pollutant load.

iodegradable: Capable of being broken down by microorganisms.

iogenesis: The production of living organisms only from other living organisms.

ioluminescence: The emission of light by living organisms.

iomass: The mass of living matter present in a specificed area.

iosphere: The zone of the earth that includes the lower atmosphere and the upper layers of soil and water.

iota: The animal, plant and microbial life characterizing a given region.

otulism: Food poisoning due to the toxin of Clostridium botulinum.

rownian motion: A peculiar dancing motion exhibited by finely divided particles and bacteria in suspension, due to bombardment by the molecules of the fluid.

udding: A form of asexual reproduction typical of yeast, in which a new cell is formed as an outgrowth from the parent cell.

uffer: Any susbstance in a fluid that tends to resist the change in pH when acid or alkali is added.

alorie: A unit of heat; the amount of heat required to raise the temperature of 1g of water by 1^0C.

apsid: The protein coat of a virus.

apsule: An envelope or slime layer surrounding the cell wall of certain microorganisms.

Carotenoid: A water insoluble pigment, usually yellow, orange or red which consi of along aliphatic polyene chain composed of isoprene units.

Catabolism: The dissimilation, or breakdown, of complex organic molecules, releasi energy. A part of the total process of metabolism. Compare anabolism.

Catalyst: Any substance that accelerates a chemical reaction but remains unalter in form and amount.

Cell: The microscopic, functionally and structurally basic unit of all living organism

Cell wall: A rigid external covering of the cytoplasmic membrane.

Chemoautotroph: An organism that obtains energy by oxidizing inorganic compoun Carbon dioxide is the so le source of carbon.

Chemolithotroph: An organism that uses in organic compounds as electron donc and relies on chemical compounds for energy.

Chemotaxis: The movement of an organism in response to a chemical stimulus.

Chemotroph: An organism that uses chemical compounds for energy. Compa phototroph.

Chlorophyll: A light trapping green pigment essential as an electron donor i photosynthesis.

Chloroplast: A cell plastid (specialized organelle) in plants and algae that contai chlorophyll pigments and functions in photosynthesis.

Chromosome: A gene-containing filamentous structure in a cell nucleus; the numb of chromosomes per cell nucleus is constant for each species.

Clone: A population of cells descended from a single cell.

Codon: A sequence of three nucleotide bases (in Mrna) that codes for an amin acid or the initiation or termination of a polypeptide chain.

Coenzyme: The non-protein of an enzyme.

Colony: A macroscopically visible growth of microorgamisms on a solid cultu medium.

Commensalism: A relationship between members of different species living i proximity (the same cultural environment) in which one organism benefit from the association but the other is not affected.

Conidium: An asexual spore that may be non-celled or many-celled and may be c many size and shapes. Also called conidiospore.

Conjugation: A mating process characterized by the temporary fusion of the matin partners and the transfer of genes. Conjugation occurs particularly in unicellular organisms.

Contamination: The entry of undersirable organisms into some material or object.

Cyst: A thick-walled dormant form of an organism which is resistant of desiccation, e.g., the cysts formed by certain bacteria such as Azotobacter or by various protozoa.

Cytochrome: One of a group of iron porphyrins that serve as reversible oxidation reduction carriers in repiration.

Cytoplasm: The living matter of a cell between the cell membrane and the uncleus.

Cytoplasmic membrane: A thin layer under the cell wall consisting mainly of phospholipids and proteins; it is responsible for the selective permeability properties of the cell. Also called plasma membrane.

Dark-field microscopy: A type of microscopic examination in which the microscopic field is dark and any objects, such as organisms, are brightly illuminated.

Deoxyribonucleic acid (DNA): The carrier or genetic information; a type of nucleic acid occurring in cells, containing phosphoric acid, D-2-deoxyribose, adenine, guanine, cytosine, and thymine.

Deoxyribose: A five carbon sugar having one oxygen atom fewer than the parent sugar, ribose; a component of DNA.

Differential stain: A procedure using a series of dye solutions or staining reagents to bring out differences in microbial cells.

Diploid: Having chromosomes in pairs the members of which are homologous; having twice the haploid number.

Disinfectant: An agent that frees from infection by killing the vegetative cells of microorganisms.

Ecology: The study of the interrelationships that exist between organisms and their environment.

Ecosystem: A functional system which includes the organisms of a natural communit together with their environment.

Endoplasmic reticulum: An extensive array of internal membranes in a eucaryotic cell.

Enzyme: An organic catalyst produced by an organism.

Epidemiology: The study of the occurrence and distribution of disease.

Eubacteria: One of the two major groups of bacteria (the other being the archaeobacteria). Eubacteria have fundamental features that are considered t be typical of most bacteria.

Eukaryote: A cell that possesses a definitive or true nucleus. Compare prokaryote

Exponential phase (log phase): The period of culture growth when cells divide steadily at a constant rate. Also called logarithmic phase.

Filter bacteriological: A special type of filter through which bacterial cells cannot pas

Fission, binary: A single nuclear division followed by the division of the cytoplasm to form two daughter cells of equal size.

Flagellum pl. flagella: A thin, filamentous appendage on cells, responsible fo swimming motility.

Fluorescence microscopy: Microbiology in which cells or their components are staine with a fluorescent dy and thus appear as glowing objects against a dark backgrounc

Fungus, pl. fungi: A microorganisms that lacks chlorophyll and is usually filamentous a mold or yeast.

Gene: A segment of a chromosome, definable in operational terms as the respositor of a unit of genetic information.

Generation time: The time interval necessary for a cell to divide.

Genome: A complete set of genetic material; i.e., a complete set of genes.

Glogi apparatus: A membranous organelle in the endoplasmic reticulum of the cell

Gram negative bacteria: Bacteria that do not retain the crystal violent-iodine complex when subjected to the Gram technique and thus acquire the color of the dye (usually a red dye) that is used to counterstain the cells.

Guanine: A purine base, occurring naturally as a fundamental component of nucleic acids.

Habitat: The natural environment of an organism.

Halophile: A microorganism whole growth is accelerated by or dependent on high salt concentrations.

Heterotroph: A microorganism that is unable to use carbon dioxide as its sole source of carbon and requires one or more organic compounds. Compare autotroph.

Holoenzyme: A fully active enzyme, containing an apoenzyme and a coenzyme.

Incubation: In microbiology, the subjecting of cultures of microorganisms to conditions (especially temperatures) favourable to their growth.

nhibition: In microbiology, the prevention of the growth or multiplication of microorganisms.

noculation: The artificial introduction of microorganisms or other substance into the body or into a culture medium.

ntercellular: Between cells.

ntracellular: Within a cell.

ag phase: The period of slow, ordely growth when a medium is first inoculated with a culture.

ipase: An enzyme that catalyzes the hydrolysis of fats into glycerol and fatty acids.

ogarithmic phase: Commonly called log phase. See exponential phase.

yophilization: The preservation of biological specimens by rapid freezing and rapid dehydration in a high vaccum.

Medium, pl. media: A substance used to provide nutrient for the growth and multiplication of microorganisms.

Meiosis: A process occurring during cell division at different points in the life cycles of different organisms, in which the chromosome number is reduced by half, thus compensating for the chromosome doubling effect of fertilization compare mitosis.

Mesophile: A bacterium growing best at the moderate temperature range 25 to 40^{0}C.

Mesosomes: Membrane invaginations in the form of convoluted tubules and vesicles.

Messenger RNA: The intermediary substance that passes information from the DNA in the nuclear region to the ribosomes in the cytoplasm. Abbreviation: mRNA.

Metabolism: The system of chemical changes by which the nutritional and functional activities of an organism are maintained.

Metabolite: Any chemical participating in metabolism; a nutrient.

Microbe: Any microscopic organism; a microorganism. Adjective: microbial.

Microbiology: The study of organisms of microscopic size (microorganisms), including their culture, economic importance, pathogenicity, etc.

Microorganism: Any organism of microscopic dimensions.

Mitochondrion: A cytoplasmic organelle in eukaryotic cells; the site of cell respiration.

Mitosis: A form of nuclear division characterized by complex chromosome movement and exact chromosome movement and exact chromosome duplication. Compare meiosis.

Mold: A fungus, characterized by a filamentous structure.

Morphology: The branch of biological science that deals with the study of th structure and form of living organisms.

Mutation: A stable change of gene such that the changed condition is inherited b offspring cells.

Mycelium: A mass of threadlike filaments, branched or composing a network, tha constitutes the vegetative structure of a fungus.

Mycology: The study of fungi.

Nomenclature: Any system of scientific names, such as those employed in biologica classificaition.

Nucleic acid: One of a class of molecules composed of joined nucleotide complexe the types are deoxyribonucleic acid (DNA) and ribonucleic acid (RNA).

Nucleolus, pl. nucleoli: A small body in a cell nucleus.

Nucleoside: A pentose sugar linked to a purine or pyrimidine base.

Nucleotide: The basic bulding block of nucleic acids (DNA and RNA); consists c a purine or pyrimidine base, ribose or deoxyribose, and phosphate.

Nucleus: The structure in a cell that contains the chromosomes.

Osmosis: The passage, due to osmotic pressure, of a fluid through a semipermeabl membrane.

Oxidation: 1. The process of combining with oxygen. 2. The loss of electrons o hydrogen atoms.

Pasteurization: The process of heating a liquid food or beverage to a controlled temperature to enhance the keeping quality and destroy harmful microorganisms

Pathogen: An organism capable of producing disease.

Periplasmic space: The space between the cytoplasmic membrance and the oute membrane of Gram-negative bacteria.

Periplast: A surface membrane or pellicle of certain algae and bacteria.

pH: A symbol for the degree of acidity or alkalinity of a solution; pH = log (1/[H+]) represents the hydrogen ion concentration.

Phosphorylation: The addition of a phosphate group to a compound.

Photolysis: Light generated break down of water.

Phototroph: A bacterium capable of utilizing light energy for metabolism.

ıylogeny: The evolutionary or ancestral history of organisms.

ıylum: A taxon consisting of a group of related classes.

ıysiology: The study of the life processes of living things.

lus: Any filamentous appendage other than flagella on certain Gram-negative bacteria.

okaryote: A type of cell in which the nuclear substance is not enclosed within a membrane; e.g., a bacterium or cyanobacterium. Compare Eukaryote.

otoplasm: The living substance of a cell. The term usually refers to the substance enclosed by the cytoplasmic membrane.

ychrotroph: an organism which is able to grow at 0^0C but which grows best at a temperature above 20^0C; also termed facultative psychrophile.

eduction: A chemical process involving the removal of oxygen, the addition of hydrogen atoms, or the gain of electrons.

espiration: An energy-yielding process in which electrons from an oxidizable substrate are transferred via a series of oxidation-reduction reactions to an exogenous terminal electron acceptor.

hizospher: The soil region subject to the influence of plant roots and characterized by a zone of increased microbiological activity.

ibonucleic acid (RNA): A nucleic acid occurring in the cytoplasm and the nucleolus, containing phosphoric acid, D-ribose, adenine, guanine, cytosine and uracil.

ibosomal RNA (rRNA): The RNA of the ribosomes constituting about 90 per cent of the total cellular RNA.

ibosome: A cytoplasmic structural unit, made up of RNA and protein that is the site of protein synthesis.

aprophyte: An organism living on dead organic matter.

pecies: The basic taxonomic group; in bacteriology, a species consists of a type strain together will all the other strains that are considered sufficiently similar to the type strain to warrant inclusion in the species.

pore: A resistant body formed by certain microorganisms.

terilization: The process of making sterile; the killing of all forms of life.

ubstrate: The substance acted upon by an enzyme.

ymbiosis: The living together of two or more organisms; microbial association.

Synergism: The ability of two or more organisms to bring about changes (usua chemical) that neigher can accomplish alone.

Taxonomy: The classification (arrangement), nomenclature (naming), and identificati of organisms.

Thermophile: An organism that grows best at temperatures above 45^{0}C.

Transcription: The process in which a complementary single stranded mRNA synthesized from one of the DNA strands.

Transduction: The transfer of genetic material from one bacterium to another throu the agency of a virus.

Transfer RNA (tRNA): A specific RNA for each amino acid that becomes esterified the terminal adenosine. Each of the 60 or so tRNA As has a specific trinucleoti sequence that interacts with a complementary sequence in mRNA. Also call soluble RNA (sRNA).

Transformation: The phenomenon by which certain bacteria incorporate DNA fro related strains into their genetic makeup.

Tyndallization: A process of fractional sterilization with flowing stream.

Ultraviolet rays: Radiations from about 3900 to about 2000A.

Vaccine: A preparation of killed or attenuated microorganisms, or their componen or their products that is used to induce active immunity against a disease.

Virion: The complete mature virus particle

Virucide: An agent that kills viruses.

Virus: An obligate intracellular parasitic microorganism that is smaller than bacteri More viruses can pass through filters that retain bacteria.

Yeast: A kind of spore resulting from the fusion of two similar gametes in some fung

Zygospore: A kind of spore resulting from the fusion of two similar gametes i some fungi.

Zygote: An organism produced by the union of two gemetes.

References

Allen, O. N. (1957) *Experiments in soil bacteriology*. 3rd ed. Burgess Publishing Co., Minneapolis, USA.

Abdel-Bari, N. A., Aboul-Eid, H. Z., Anter, E. A. and Noweer, E. A. Effects of different fungal filtrates on Meloidogyne incognita larvae in laboratory bioassay tests. Egyptian J. Agronematol. 4: 49–69.2000

Add-Elmoity, Riad, F. W. and El-Eraki, S. Effect of single mixture of antagonistic fungi on the control of root-knot nematode Meloidogyne incognita. Egyptian J. Agric. 71: 91–101.1993

Al-Fattah, A., Dababat, A. and Sikora, R. A. Use of Trichoderma harzianum and Trichoderma viride for the biological control of Meloidogyne incognita on tomato. Jordan J. Agric. Sci. 3 (3): 297–309.2007

Ali, A. H. H. and Barakat, M. I. E. Utilisation of Trichoderma harzianum as a biocontrol agent against root-knot nematode Meloidogyne incognita. Egyptian J. Biol. Pest Control. 4: 67–77.1994

Ausmus, B. S. and O'Neill, E. G. (1978) "Comparison of carbon dynamics of three microcosm substrates." *Soil Biol. Biochem*. 10: 425-429.

Beaucham, E.G. and Hume, D.J. (1997) "Microorganisms to enhance nitrogen fixation." In: *Modern Soil Microbiology* (eds.) Van Elasa, J.D., Trevres, J.T., Wellington, E.M.H. Pg. 658-659.

Beijerinck, M. W. (1901). Uber oligonitrophile Mikroben. Zbl. Backt. 7, 561-582

Benson Harold, J. (1990) "Microbiological Applications", *A Laboratory Manual in General Microbiology*, Part I Microscopy. Pg. 22-24.

Badr, S. T. A.: Effects of seven fungal filtrates, singly and combined with three nematicides on Meloidogyne javanica juveniles. Egyptian J. Agronematol. 5: 105–113.2001

Baker, R. and Griffin, G. J. Molecular strategies for biological control of fungal plant pathogens. In R. Reuveni (ed.), Novel Approaches to Integrated Pest Management, pp. 153–82. Boca Raton, FL, USA: Lewis Publisher. 1995

Bokhari, F. M. Efficacy of some Trichoderma species in the control of Rotylenchulus reniformis and Meloidogyne javanica. Archives Phytopathol. and Plant Prot. 42 (4): 361–369.2009.

Boland, G. J. Biological control of plant diseases with fungal antagonists: Challenges an opportunities. Canadian J. Plant Pathol. 12: 295–299.1990

Case L. Christine and Johnson Ted R. (1984) *Laboratory Experiments in Microbiolog* Benjamin Cummings Publishing Company, Inc., Menlo Park, California. Pg. 15-16

Cobb, N. A. (1918) "Estimating nematode population of soil."U.S. Dep. Agric. *Bu Plant Ind. Agric. Tech. Circ.* 1: 1-48.

Christie, J. R. and Perry, V. G. (1951) "Removing nematodes from soil." *Pro Helminthol. Soc. Wash.* 18: 106-108.

Chet, I. Trichoderma - application, mode of action and potential as biocontrol agent of soi borne plant pathogenic fungi. In: I Chet (ed.), Innovative Approaches to Plant Diseas Control, pp. 137-160. New York: John Wiley & Sons. 1987

Deka, S. B. K. Studies on nematodes associated with mulberry, Morus alba L. M.Sc. Thesis TNAU, Coimbatore, Tamil Nadu, India.1994

Dubey, S. C., Suresh, M. and Singh, B. Evaluation of Trichoderma species against Fusariun oxysporum fsp. circeris for integrated management of chick pea wilt. Biological Contro 40: 118–127.2007

Duffy, B. K., Simon, A. and Weller, D. M. Combination of Trichoderma koningii witl fluorescent pseudomonads for control of take-all on wheat. Phytopathol. 86: 188-194.1996

Diaz, R. J. & Rosenberg, R. (2008). "Spreading dead zones and consequences fo marine ecosystems." Science 321: 926-929.

Difco Manual Dehydrated Culture Media and Reagents for Microbiology (1984) Difco Laboratories Detroit Michigan (USA).

Doetsch, R.N. (1981) "*Determinative Methods of Light Microscopy.*" In: *Manual of Method for General Bacteriology,* Washington.

Fred, E. B., Baldwin, J.L. and McCoy, E. (1932). *Root nodule bacteria and leguminou plants.* University of Wisconsin Studies, Madison Science, 5: 343-349.

Elad, Y., Zimmand, G., Zags, Y., Zuriel, S. and Chet, I. Use of Trichoderma harzianum in combination or alteration with fungicides to control cucumber grey mold (Botrytis cinerea) under commercial greenhouse condition. Plant Pathology. 42: 324–356.1993

Elad, Y. Biological control of foliar pathogens by means of Trichoderma harzianum and potential modes of action. Crop Prot. 19: 709–714.2000

Farouk, M. I., Rahman, M. L. and Bari, M.A. Management of root - knot nematode of tomato using Trichoderma harzianum and organic soil amendment. Bangladesh J. Plant Pathology. 18: 33–37.2002

. Freeman, S., Minz, D., Kolesnik, I., Barbul, O., Zreibil, A., Maymon, M., Nitzani, Y.,

Kirshner, B., Rav-David, D., Bitu, A., Dag, A., Shafir, S. and Elad, Y. Trichoderma biocontrol of Collectotrichum acutatum and Botrytis cinerea and survival in strawberry. European J. Plant Pathology. 110: 361–370.2004

Gerhardt, P., R.G.E. Murray, R.N. Costilow, E.W. Nester, W.A. Wood, N.R. Kreig and G. B. Phillips (1981) "Manual of Methods for General Bacteriology," Washington: *American Society for Microbiology*

Ghazalibiglar H. Hampton J.G., de Jong E.V.Z., Holyoake (2016) "Is induced systemic resistance the mechanism for control of black rot in *Brassica oleracea* by a Paenibacillus sp."? *Biol Control* 92:195–201

Harley, J. L. and Waid, J. W. (1955) A method for studying active mycelia on living roots and other surface in the soil. *Trans Br. Mycol. Soc.* 38: 104-118.

Hallmann, J. and Sikora, R. A. Toxicity of fungal endophyte secondary metabolites to plant parasitic nematodes and soil-borne plant pathogenic fungi. European J. Pl. Pathol. 102: 155–162.1996

Harman, E.G. (1991): Seed treatments for biological control of plant diseases. Crop Protection. 166 – 171.

Inbar, J., Abramsky, M., Cohen, D. and Chet, I. Plant growth enhancement and disease control by Trichoderma harzianum in vegetable seedlings grown under commercial conditions. European J. Plant Pathol. 100: 337–346.1994

Kerry, B. R. Biological control. In R. H. Brown and B. R. Kerry (eds.), Principles and Practices of Nematode Control in Crops, pp. 233–263. New York: Academic Press.1987

Kanwar, S.S., Gupta, M.K. and Punj, V. (1997) *Laboratory Manual of Basic Microbiology*, Department of Microbiology, College of Basic Sciences, Himachal Pradesh Krishi Vishvavidyalaya, Palampur.

Murray, R.G.E. and C.F. Robinow (1981). Manual of Methods for General Bacteriology, Washington. *American Society for Microbiology.*

Nonomura, H., and Y. Ohara (1969). Distribution of actinomycetes in soil. IV. A culture method effective for both preferential isolation and enumeration of *Microbispora* and *Streptosporangium* strains in soil (Part I). Hakku Kogaku Zasshi 49: 895-903.

Oostendrop, M. and Sikora, R. A. Utilisation of antagonistic rhizobacteria as a seed treatment for the biological control of Heterodera schachtii in sugarbeet. Revue de Nematol. 12: 77–83.1989

Orole, O. O. and Adejumo, T. O. Activity of fungal endophytes against four maize wilt pathogens. African J. Microbiol. 3(1): 969–973.2009

Pandey, G., Pandey, R. K. and Paul, H. Efficacy of different levels of Trichoderma viride against root-knot nematode in chickpea (Cicer arientinum L.). Annu Plant Protect. 26: 971–977.2003

Papavizas, G. C. Trichoderma and Gliocladium biology, ecology, and potential for biocontr Annual Review Phytopathol. 23: 23–54.1985

Parvatha, R. P., Rao, M. S. and Nagesh, M. Management of citrus nematode, Tylenchul semipenetrans by integration of Trichoderma harzianum with oil cakes. Nemat Medit. 24: 265-267.1996

Rao, V. R. and Swarup, G. Pathogenicity of the spiral nematode, Helicotylenchus dihyste to sugarcane. Indian J. Nematol. 4 (2):160–166.1974

Reddy, P. P., Rao M. S. and Nagesh, M. Management of the citrus nematode, Tylenchul semipenetrans by integration of Trichoderma harzianum with oil cakes. Nematolog Mediterranea. 24: 265–267.1996

Sharon, E., Bar-Eyal, M., Chet, I., Herrera-Etrella, A., Kleifeld, O. and Spiegel, Y. Biologic control of the root-knot nematode Meloidogyne javanica by Trichoderma harzianur Phytopathology. 91 (7): 687–693.2001

Sharon, E., Chet, I., Viterbo, A., Bar-Eyal, M., Nagan, H. and Samuels, G. J. Parasitism Trichoderma on Meloidogyne javanica and role of the gelatinous matrix. European Plant Pathol. 118: 247–258.2007

Sikora, R. A., Niere, B. and Kimenju, J. Endophytic microbial biodiversity and plant nematoc management in African agriculture. In P. Neuenschwander, C. Borgermeister and Langewalder (eds.), Biological Control in IPM Systems in Africa, pp. 179–192.200

Sivan, A. and Chet, I. Microbial control of plant diseases. In R. Mitchell (ed.), New Concep in Environmental Microbiology, pp. 335-354. New York: Wiley-Liss Inc.1992

. Spiegel, Y. and Chet, I. Evaluation of Trichoderma spp. as a biocontrol agent against soi borne fungi and plant-parasitic nematodes in Israel. Integrated Pest Managemer Reviews. 3: 169–175.1998

Saxena, N.P. and Awasthi, D.K. (1985) *Krishna's Microbiology*. Krishna Prakashan Medi (P) Ltd., Meerut. Pg. 141-178.

Schlegal Hans G. (1986) *General Microbiology*. Cambridge University Press. Pg. 355-358.

Somani, L.L. (2005) *Handbook of Biofertilizers*. Agrotech Publishing Academy, Udaipu Pg. 43-178.

Spencer, M. (1982) *Fundamentals of Light Microscopy*. New York: Cambridge Universit Press.

Singh J.S., Pandey V.C., Singh D.P. (2011). "Efficient soil microorganisms: A nev dimension for sustainable agriculture and environmental development". *Agr Ecosyst. Environ.* 140(3):339–353.

Starr, M. P., H. Stolp, H. G. Truper, A. Balows, and H.G. Schlegel (Eds.) (1981) *The Prokaryotes. A handbook on habitats, isolation and identification of bacteria* Vol. II. Springer-Verlag, New York.

ıbba Rao, N.S. (1993). *Biofertilizers in agriculture and forestry.* Oxford and IBH Publishing Co. Pvt. Ltd., Pg. 70-73.

an Gastel, M. Ladd, S. N. and Amato, M. (1990) "Carbon and nitrogen mineralization from two soils of contrasting texture and microaggregate stability. Influence of sequential fumigation drying and storage." *Soil Biology and Biochemistry.* 23: 313-322 (Soil respiration-C mineralization).

/ollum II A G (1982) "Cultural methods for soil microorganisms." In A L Page, R H Miller and D R. Keeney (eds.). *Methods of Soil Analysis, Part II. Chemical and Microbiological properties.* Agron. Monograph. 9 (Second Ed): 781-802.

/aksman, S. A. and Fred, B. A tentative outline of the plate method for determining the number of micro-organisms in the soil. Soil Sciences. 14: 27–28.1922

Vindham, G. L., Windham, M. T. and Williams, W. P. Effects of Trichoderma spp. on maize growth and Meloidogyne arenaria reproduction. Plant Des. 73: 493–494.1989

